Python Real-World Projects

Craft your Python portfolio with deployable applications

Steven F. Lott

BIRMINGHAM—MUMBAI

"Python" and the Python logo are trademarks of the Python Software Foundation.

Python Real-World Projects

Associate Group Product Manager: Kunal Sawant
Publishing Product Manager: Akash Sharma
Senior Editor: Kinnari Chohan
Senior Content Development Editor: Rosal Colaco
Technical Editor: Maran Fernandes
Copy Editor: Safis Editing
Associate Project Manager: Deeksha Thakkar
Proofreader: Safis Editing
Indexer: Pratik Shirodkar
Production Designer: Shyam Sundar Korumilli
Business Development Executive: Debadrita Chatterjee
Developer Relations Marketing Executive: Sonia Chauhan

First published: September 2023

Production reference: 1210823

Published by Packt Publishing Ltd.
Grosvenor House
11 St Paul's Square
Birmingham
B3 1RB

ISBN 978-1-80324-676-5

www.packtpub.com

Contributors

About the author

Steven F. Lott has been programming since computers were large, expensive, and rare. Working for decades in high tech has given him exposure to a lot of ideas and techniques; some are bad, but most are useful and helpful to others.

Steven has been working with Python since the '90s, building a variety of tools and applications. He's written a number of titles for Packt Publishing, including *Mastering Object-Oriented Python*, *Modern Python Cookbook*, and *Functional Python Programming*.

He's a tech nomad and lives on a boat that's usually located on the east coast of the US. He tries to live by the words, "Don't come home until you have a story."

About the reviewer

Chris Griffith is a Lead Software Engineer with twelve years of experience with Python. His open-source Python projects have been downloaded over a million times, and he is the primary writer for the **Code Calamity** blog. Chris enjoys studio photography in his free time as well as digitizing vintage magazines and 8mm films.

Join our community Discord space

Join our Python Discord workspace to discuss and learn more about the book:

https://packt.link/dHrHU

Table of Contents

Chapter 15: Project 5.1: Modeling Base Application 379

Chapter 16: Project 5.2: Simple Multivariate Statistics 405

Preface

How do we improve our knowledge of Python? Perhaps a more important question is "How do we show others how well we can write software in Python?"

Both of these questions have the same answer. We build our skills and demonstrate those skills by completing projects. More specifically, we need to complete projects that meet some widely-accepted standards for professional development. To be seen as professionals, we need to step beyond apprentice-level exercises, and demonstrate our ability to work without the hand-holding of a master crafter.

I think of it as sailing a boat alone for the first time, without a more experienced skipper or teacher on board. I think of it as completing a pair of hand-knitted socks that can be worn until the socks have worn out so completely, they can no longer be repaired.

Completing a project entails meeting a number of objectives. One of the most important is posting it to a public repository like SourceForge (`https://sourceforge.net`) or GitHub (`https://github.com`) so it can be seen by potential employers, funding sources, or business partners.

We'll distinguish between three audiences for a completed project:

- A personal project, possibly suitable for a work group or a few peers.

- A project suitable for use throughout an enterprise (e.g., a business, organization, or government agency)

- A project that can be published on the Python Package Index, PyPI (`https://pypi.org`).

We're drawing a fine line between creating a PyPI package and creating a package usable within an enterprise. For PyPI, the software package must be installable with the **PIP** tool; this often adds requirements for a great deal of testing to confirm the package will work in the widest variety of contexts. This can be an onerous burden.

For this book, we suggest following practices often used for "Enterprise" software. In an Enterprise context, it's often acceptable to create packages that are *not* installed by **PIP**. Instead, users can install the package by cloning the repository. When people work for a common enterprise, cloning packages permits users to make pull requests with suggested changes or bug fixes. The number of distinct environments in which the software is used may be very small. This reduces the burden of comprehensive testing; the community of potential users for enterprise software is smaller than a package offered to the world via PyPI.

Who this book is for

This book is for experienced programmers who want to improve their skills by completing professional-level Python projects. It's also for developers who need to display their skills by demonstrating a portfolio of work.

This is not intended as a tutorial on Python. This book assumes some familiarity with the language and the standard library. For a foundational introduction to Python, consider *Learn Python Programming, Third Edition*: `https://www.packtpub.com/product/learn-python-programming-third-edition/9781801815093`.

The projects in this book are described in broad strokes, requiring you to fill in the design details and complete the programming. Each chapter focuses more time on the desired approach and deliverables than the code you'll need to write. The book will detail test cases and acceptance criteria, leaving you free to complete the working example that passes the suggested tests.

What this book covers

We can decompose this book into five general topics:

- We'll start with **Acquiring Data From Sources**. The first six projects will cover projects to acquire data for analytic processing from a variety of sources.

- Once we have data, we often need to **Inspect and Survey**. The next five projects look at some ways to inspect data to make sure it's usable, and diagnose odd problems, outliers, and exceptions.

- The general analytics pipeline moves on to **Cleaning, Converting, and Normalizing**. There are eight projects that tackle these closely-related problems.

- The useful results begin with **Presenting Summaries**. There's a lot of variability here, so we'll only present two project ideas. In many cases, you will want to provide their own, unique solutions to presenting the data they've gathered.

- This book winds up with two small projects covering some basics of **Statistical Modeling**. In some organizations, this may be the start of more sophisticated data science and machine learning applications. We encourage you to continue your study of Python applications in the data science realm.

The first part has two preliminary chapters to help define what the deliverables are and what the broad sweep of the projects will include. *Chapter 1, Project Zero: A Template for Other Projects* is a baseline project. The functionality is a "Hello, World!" application. However, the additional infrastructure of unit tests, acceptance tests, and the use of a tool like **tox** or **nox** to execute the tests is the focus.

The next chapter, *Chapter 2, Overview of the Projects*, shows the general approach this book will follow. This will present the flow of data from acquisition through cleaning to analysis and reporting. This chapter decomposes the large problem of "data analytics" into a number of smaller problems that can be solved in isolation.

The sequence of chapters starting with *Chapter 3, Project 1.1: Data Acquisition Base Application*, builds a number of distinct data acquisition applications. This sequence starts with acquiring data from CSV files. The first variation, in *Chapter 4, Data Acquisition Features: Web APIs and Scraping*, looks at ways to get data from web pages.

The next two projects are combined into *Chapter 5, Data Acquisition Features: SQL Database*. This chapter builds an example SQL database, and then extracts data from it. The example database lets us explore enterprise database management concepts to more fully understand some of the complexities of working with relational data.

Once data has been acquired, the projects transition to data inspection. *Chapter 6, Project 2.1: Data Inspection Notebook* creates an initial inspection notebook. In *Chapter 7, Data Inspection Features*, a series of projects add features to the basic inspection notebook for different categories of data.

This topic finishes with the *Chapter 8, Project 2.5: Schema and Metadata* project to create a formal schema for a data source and for the acquired data. The JSON Schema standard is used because it seems to be easily adapted to enterprise data processing. This schema formalization will become part of later projects.

The third topic — cleaning — starts with *Chapter 9, Project 3.1: Data Cleaning Base Application*. This is the base application to clean the acquired data. This introduces the **Pydantic** package as a way to provide explicit data validation rules.

Chapter 10, Data Cleaning Features has a number of projects to add features to the core data cleaning application. Many of the example datasets in the previous chapters provide very clean data; this makes the chapter seem like needless over-engineering. It can help if you extract sample data and then manually corrupt it so that you have examples of invalid and valid data.

In *Chapter 11, Project 3.7: Interim Data Persistence*, we'll look at saving the cleaned data for further use.

The acquire-and-clean pipeline is often packaged as a web service. In *Chapter 12, Project 3.8: Integrated Data Acquisition Web Service*, we'll create a web server to offer the cleaned data for subsequent processing. This kind of web services wrapper around a long-running acquire-and-clean process presents a number of interesting design problems.

The next topic is the analysis of the data. In *Chapter 13, Project 4.1: Visual Analysis Techniques* we'll look at ways to produce reports, charts, and graphs using the power of **JupyterLab**.

In many organizations, data analysis may lead to a formal document, or report, showing the results. This may have a large audience of stakeholders and decision-makers. In *Chapter 14, Project 4.2: Creating Reports* we'll look at ways to produce elegant reports from the raw data using computations in a **JupyterLab** notebook.

The final topic is statistical modeling. This starts with *Chapter 15, Project 5.1: Modeling Base Application* to create an application that embodies lessons learned in the **Inspection Notebook** and **Analysis Notebook** projects. Sometimes we can share Python programming among these projects. In other cases, however, we can only share the lessons learned; as our understanding evolves, we often change data structures and apply other optimizations making it difficult to simply share a function or class definition.

In *Chapter 16, Project 5.2: Simple Multivariate Statistics*, we expand on univariate modeling to add multivariate statistics. This modeling is kept simple to emphasize foundational design and architectural details. If you're interested in more advanced statistics, we suggest building the basic application project, getting it to work, and then adding more sophisticated modeling to an already-working baseline project.

The final chapter, *Chapter 17, Next Steps*, provides some pointers for more sophisticated applications. In many cases, a project evolves from exploration to monitoring and maintenance. There will be a long tail where the model continues to be confirmed and refined. In some cases, the long tail ends when a model is replaced. Seeing this long tail can help an analyst understand the value of time invested in creating robust, reliable software at each stage of their journey.

A note on skills required

These projects demand a wide variety of skills, including software and data architecture, design, Python programming, test design, and even documentation writing. This breadth of skills reflects the author's experience in enterprise software development. Developers are expected to be generalists, able to follow technology changes and adapt to new technology.

In some of the earlier chapters, we'll offer some guidance on software design and construction. The guidance will assume a working knowledge of Python. It will point you toward the documentation for various Python packages for more information.

We'll also offer some details on how best to construct unit tests and acceptance tests. These topics can be challenging because testing is often under-emphasized. Developers fresh out of school often lament that modern computer science education doesn't seem to cover testing and test design very thoroughly.

This book will emphasize using **pytest** for unit tests and **behave** for acceptance tests. Using **behave** means writing test scenarios in the Gherkin language. This is the language used by the **cucumber** tool and sometimes the language is also called Cucumber. This may be new, and we'll emphasize this with more detailed examples, particularly in the first five chapters.

Some of the projects will implement statistical algorithms. We'll use notation like $\bar{x}$ to represent the mean of the variable x. For more information on basic statistics for data analytics, see *Statistics for Data Science*:

`https://www.packtpub.com/product/statistics-for-data-science/9781788290678`

To get the most out of this book

This book presumes some familiarity with Python 3 and the general concept of application development. Because a project is a complete unit of work, it will go beyond the Python programming language. This book will often challenge you to learn more about specific Python tools and packages, including **pytest**, **mypy**, **tox**, and many others.

Most of these projects use **exploratory data analysis (EDA)** as a problem domain to show the value of functional programming. Some familiarity with basic probability and statistics will help with this. There are only a few examples that move into more serious data science.

Python 3.11 is expected. For data science purposes, it's often helpful to start with the **conda** tool to create and manage virtual environments. It's not required, however, and you should be able to use any available Python.

Additional packages are generally installed with `pip`. The command looks like this:

```
% python -m pip install pytext mypy tox beautifulsoup4
```

Complete the extras

Each chapter includes a number of "extras" that help you to extend the concepts in the chapter. The extra projects often explore design alternatives and generally lead you to create additional, more complete solutions to the given problem.

In many cases, the extras section will need even more unit test cases to confirm they actually solve the problem. Expanding the core test cases of the chapter to include the extra features is an important software development skill.

Download the example code files

The code bundle for the book is hosted on GitHub at `https://github.com/PacktPubl ishing/Python-Real-World-Projects`. We also have other code bundles from our rich catalog of books and videos available at `https://github.com/PacktPublishing/`. Check them out!

Conventions used

There are a number of text conventions used throughout this book.

`CodeInText`: Indicates code words in the text, database table names, folder names, filenames, file extensions, pathnames, dummy URLs, user input, and Twitter handles. For example: "Python has other statements, such as `global` or `nonlocal`, which modify the rules for variables in a particular namespace."

Bold: Indicates a new term, an important word, or words you see on the screen, such as in menus or dialog boxes. For example: "The **base case** states that the sum of a zero-length sequence is 0. The **recursive case** states that the sum of a sequence is the first value plus the sum of the rest of the sequence."

A block of code is set as follows:

```
print("Hello, World!")
```

Any command-line input or output is written as follows:

```
% conda create -n functional3 python=3.10
```

 Warnings or important notes appear like this.

 Tips and tricks appear like this.

Get in touch

Feedback from our readers is always welcome.

General feedback: Email feedback@packtpub.com, and mention the book's title in the subject of your message. If you have questions about any aspect of this book, please email us at questions@packtpub.com.

Errata: Although we have taken every care to ensure the accuracy of our content, mistakes do happen. If you have found a mistake in this book we would be grateful if you would report this to us. Please visit https://subscription.packtpub.com/help, click on the **Submit Errata** button, search for your book, and enter the details.

Piracy: If you come across any illegal copies of our works in any form on the Internet, we would be grateful if you would provide us with the location address or website name. Please contact us at copyright@packtpub.com with a link to the material.

If you are interested in becoming an author: If there is a topic that you have expertise in and you are interested in either writing or contributing to a book, please visit `http://authors.packtpub.com`.

Share your thoughts

Once you've read *Python Real-World Projects*, we'd love to hear your thoughts! Scan the QR code below to go straight to the Amazon review page for this book and share your feedback.

`https://packt.link/r/1803246766`

Your review is important to us and the tech community and will help us make sure we're delivering excellent quality content.

Download a free PDF copy of this book

Thanks for purchasing this book!

Do you like to read on the go but are unable to carry your print books everywhere? Is your eBook purchase not compatible with the device of your choice?

Don't worry, now with every Packt book, you get a DRM-free PDF version of that book at no cost.

Read anywhere, any place, on any device. Search, copy, and paste code from your favorite technical books directly into your application.

The perks don't stop there, you can get exclusive access to discounts, newsletters, and great free content in your inbox daily

Follow these simple steps to get the benefits:

1. Scan the QR code or visit the link below

https://packt.link/free-ebook/9781803246765

2. Submit your proof of purchase

3. That's it! We'll send your free PDF and other benefits to your email directly

1

Project Zero: A Template for Other Projects

This is a book of projects. To make each project a good portfolio piece, we'll treat each project as an enterprise software product. You can build something that could be posted to a company's (or organization's) internal repository.

For this book, we'll define some standards that will apply to all of these projects. The standards will identify deliverables as a combination of files, modules, applications, notebooks, and documentation files. While each enterprise is unique, the standards described here are consistent with my experience as a consultant with a variety of enterprises.

We want to draw an informal boundary to avoid some of the steps required to post to the PyPI website. Our emphasis is on a product with test cases and enough documentation to explain what it does. We don't want to go all the way to creating a project in PyPI. This allows us to avoid the complications of a build system and the associated `pyproject.toml` file.

These projects are not intended to produce generic, reusable modules. They're applications specific to a problem domain and a dataset. While these are specific solutions, we don't want to discourage anyone who feels motivated to generalize a project into something generic and reusable.

This chapter will show the general outline of each project. Then we'll look at the set of deliverables. This chapter ends with project zero – an initial project that will serve as a template for others. We'll cover the following topics:

- An overview of the software quality principles that we'll try to emphasize.

- A suggested approach to completing the project as a sequence of project sprints.

- A general overview of the list of deliverables for each project.

- Some suggested tools. These aren't required, and some readers may have other choices.

- A sample project to act as a template for subsequent projects.

We'll start with an overview of some characteristics of high-quality software. The idea is to establish some standards for the deliverables of each project.

On quality

It helps to have a clear definition of expectations. For these expectations, we'll rely on the ISO 25010 standard to define quality goals for each project. For more details, see `https://iso25000.com/index.php/en/iso-25000-standards/iso-25010`.

The ISO/IEC 25010:2011 standard describes **Systems and software Quality Requirements and Evaluation (SQuaRE)**. This standard provides eight characteristics of software. These characteristics are as follows:

- **Functional suitability**. Does it do what we need? It is complete, correct, and appropriate for the user's expressed (and implied) needs? This is the focus of each project's description.

- **Performance efficiency**. Does it work quickly? Does it use the minimum resources? Does it have enough capacity to meet the user's needs? We won't address this deeply in this book. We'll talk about writing performance tests and ways to address performance concerns.

- **Compatibility**. Does it co-exist with other software? Does it properly interoperate with other applications? To an extent, Python can help assure an application interoperates politely with other applications. We'll emphasize this compatibility issue in our choices of file formats and communication protocols.

- **Usability**. There are a number of sub-characteristics that help us understand usability. Many of the projects in this book focus on the **command-line interface (CLI)** to assure a bare minimum of learnability, operability, error protection, and accessibility. A few projects will include a web services API, and others will make use of the GUI interface of JupyterLab to provide interactive processing.

- **Reliability**. Is it available when the users want it? Can we detect and repair problems? We need to make sure we have all of the parts and pieces so we can use the software. We also need to make sure we have a complete set of tests to confirm that it will work.

- **Security**. As with usability, this is a deep topic. We'll address some aspects of security in one of the projects. The remaining projects will use a CLI permitting us to rely on the operating system's security model.

- **Maintainability**. Can we diagnose problems? Can we extend it? We'll look at documentation and test cases as essential for maintainability. We'll also leverage a few additional project files to make sure our project can be downloaded and extended by others.

- **Portability**. Can we move to a new Python version? New hardware? This is very important. The Python ecosystem is rapidly evolving. Since all of the libraries and packages are in a constant state of change, we need to be able to define precisely what packages our project depends on, and confirm that it works with a new candidate set of packages.

Two of these characteristics (Compatibility and Portability) are features of Python. A wise choice of interfaces assures that these characteristics are met. These are sometimes described as architectural decisions since they influence how multiple applications work together.

For Security, we will rely on the operating system. Similarly, for Usability, we'll limit ourselves to CLI applications, relying on long-standing design principles.

The idea of Performance is something we won't emphasize here. We will point out places where large data sets will require some careful design. The choice of data structure and algorithm is a separate subject area. Our objective in this book is to expose you to projects that can provide the stimulus for a deeper study of performance issues.

Three of these quality characteristics — Functional suitability, Reliability, and Maintainability — are the real focus of these projects. These seem to be essential elements of good software design. These are the places where you can demonstrate your Python programming skills.

Another view is available from **The Twelve-Factor App** (`https://12factor.net`). This is narrowly focused on web applications. The concepts provide deeper insights and more concrete technical guidance into the quality characteristics shown above:

I. Codebase. "One codebase tracked in revision control, many deploys." We'll use **Git** and **GitHub** or perhaps one of the other version managers supported by **sourceforge**.

II. Dependencies. "Explicitly declare and isolate dependencies." Traditionally, a Python `requirements.txt` file was used for this. In this book, we'll move forward to using a `pyproject.toml` file.

III. Config. "Store config in the environment." We won't emphasize this, but Python offers numerous ways to handle configuration files.

IV. Backing services. "Treat backing services as attached resources." We touch on this in a few places. How storage, messages, mail, or caching work isn't something we'll examine deeply.

V. Build, release, run. "Strictly separate build and run stages." For command-line applications, this means we should deploy the application into a "production" environment to use the high-value data and produce the results that the enterprise needs. We want to avoid running things in our desktop development environment.

VI. Processes. "Execute the app as one or more stateless processes." CLI applications tend to be structured this way without making any additional effort.

VII. Port binding. "Export services via port binding." We won't emphasize this; it's very specific to web services.

VIII. Concurrency. "Scale out via the process model." This is a subject for the interested reader who wants to process very large data sets. We won't emphasize it in the main text. We will suggest some of these topics in the "Extras" section of some chapters.

IX. Disposability. "Maximize robustness with fast startup and graceful shutdown." CLI applications tend to be structured this way, also.

X. Dev/prod parity. "Keep development, staging, and production as similar as possible." While we won't emphasize this deeply, our intent with CLI applications is to expose the distinctions between development and production with command-line arguments, shell environment variables, and configuration files.

XI. Logs. "Treat logs as event streams." We will suggest applications write logs, but we won't provide more detailed guidance in this book.

XII. Admin processes. "Run admin/management tasks as one-off processes." A few of the projects will require some additional administrative programming. These will be built as deliverable CLI applications, complete with an acceptance test suite.

Our objective is to provide project descriptions and lists of deliverables that try to conform to these quality standards. As we noted earlier, each enterprise is unique, and some organizations will fall short of these standards, while some will exceed them.

More Reading on Quality

In addition to the ISO standard, the IEEE 1061 standard also covers software quality. While it has been inactive since 2020, it contains some good ideas. The standard is focused on quality *metrics*, which dives deeply into the idea of analyzing software for quality factors.

It can also help to read `https://en.wikipedia.org/wiki/ISO/IEC_9126` for some background on the origins of the ISO standard.

When doing more reading on this topic, it can help to recognize the following three terms:

- **Factors** are an external view of the software. They reflect the user's understanding. Some of the underlying quality characteristics are not directly visible to users. Maintainability, for example, may appear to users as a reliability or usability problem because the software is difficult to repair or extend.

- **Criteria** come from an internal view of the software. Quality criteria are the focus of the project's deliverables. Our project code should reflect the eight quality characteristics listed above.

- **Metrics** are how we can control the factors that are seen by the user. We won't emphasize quality metrics. In some cases, tools like **pylint** provide tangible measurements of static code quality. This isn't a comprehensive tool for software quality in general, but it provides an easy starting point for a few key metrics related to complexity and maintainability.

Given these standards for high-quality software, we can turn our attention to the sequence of steps for building these files. We'll suggest a sequence of stages you can follow.

Suggested project sprints

We hesitate to provide a detailed step-by-step process for building software. For more experienced developers, our sequence of steps may not match their current practices. For less experienced developers, the suggested process can help by providing a rational order in which the deliverables can be built.

There was a time when a "statement of work" with a detailed list of specific tasks was a central part of a software development effort. This was often part of a "waterfall" methodology where requirements flowed to analysts who wrote specifications that flowed down to designers who wrote high-level designs that flowed down to coders. This wasn't a great way to build software, and has been largely supplanted by Agile methods. For more information on Agility, see `https://agilemanifesto.org`.

The Agile approach lets us examine a project both as a series of steps to be completed, as well as a collection of deliverables that need to be created. We'll describe the steps first, avoiding too much emphasis on details. We'll revisit the deliverables, and in those sections, dive a little more deeply into what the final product needs to be.

The suggested approach follows the "Agile Unified Process" (`https://www.methodsandtools.com/archive/archive.php?id=21`), which has four general phases. We'll subdivide one of the phases to distinguish two important kinds of deliverables.

We suggest tackling each project in the following five phases:

1. Inception. Ready the tools. Organize the project directory and virtual environment.

2. Elaboration, part 1: Define done. This is implemented as acceptance test cases.

3. Elaboration, part 2: Define components and some tests. This is implemented as unit test cases for components that need to be built.

4. Construction. Build the software.

5. Transition. Final cleanup: make sure all tests pass and the documentation is readable.

These efforts don't proceed in a simple linear fashion. It's often necessary to iterate between elaboration and construction to create features separately.

It often works as shown in *Figure 1.1*.

This figure provides a very coarse overview of the kinds of activities we'll discuss below. The important concept is iterating between the elaboration and construction phases. It's difficult to fully design a project before constructing all of the code. It's easier to design a little, construct a little, and refactor as needed.

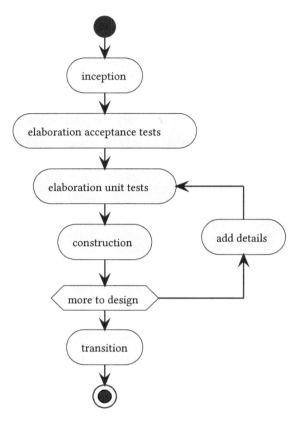

Figure 1.1: Development Phases and Cycles

For a complex project, there may be a series of transitions to production. Often a "minimally viable product" will be created to demonstrate some of the concepts. This will be followed by products with more features or features better focused on the user. Ideally, it will have both kinds of enhancements: more features and a better focus on the user's needs.

We'll look at each of these four phases in a little more detail, starting with the inception phases.

Inception

Start the inception phase by creating the parent directory for the project, then some commonly-used sub-directories (docs, notebooks, src, tests). There will be some top-level files (`README.md`, `pyproject.toml`, and `tox.ini`). The list of expected directories and files is described in more detail in *List of deliverables*, later in this chapter. We'll look at the contents of each of these files and directories in the *Deliverables* section.

It helps to capture any initial ideas in the `README.md` file. Later, this will be refactored into more formal documentation. Initially, it's the perfect place to keep notes and reminders.

Build a fresh, new virtual environment for the project. Each project should have its own virtual environment. Environments are essentially free: it's best to build them to reflect any unique aspects of each project.

Here's a **conda** command that can be used to build an environment.

```
% conda create -n project0 --channel=conda-forge python=3.10
```

An important part of inception is to start the documentation for the project. This can be done using the Sphinx tool.

While Sphinx is available from the **Conda Forge**, this version lags behind the version available from the **PyPI** repository. Because of this lag, it's best to install Sphinx using **PIP**:

```
% python -m pip install sphinx
```

After installing Sphinx, it helps to initialize and publish the documentation for the project. Starting this permits publishing and sharing the design ideas as the work progresses. In the `docs` directory, do the following steps:

1. Run the `sphinx-quickstart` command to populate the documentation. See `https://www.sphinx-doc.org/en/master/usage/quickstart.html#setting-up-the-documentation-sources`.

2. Update the `index.rst` table of contents (TOC) with two entries: "overview" and "API". These are sections that will be in separate files.

3. Write an `overview.rst` document with the definition of done: what will be accomplished. This should cover the core "Who-What-When-Where-Why" of the project.

4. Put a title in the API document, and a `.. todo::` note to yourself. You'll add to this document as you add modules to your project.

5. During Elaboration, you'll update the the `index.rst` to add sections for architecture and design decisions.

6. During Construction, as you create code, you'll add to the API section.

7. During Transition, you'll add to the `index.rst` with some "How" sections: How to test it, and how to use it.

With this as the starting point, the `make html` command will build a documentation set in HTML. This can be shared with stakeholders to assure there's a clear, common understanding of the project.

With a skeleton directory and some initial places to record ideas and decisions, it makes sense to start elaborating on the initial goal to and decide what will be built, and how it will work.

Elaboration, part 1: define done

It helps to have a clear definition of "done." This guides the construction effort toward a well-defined goal. It helps to have the definition of done written out as a formal, automated test suite. For this, the Gherkin language is helpful. The **behave** tool can execute the Gherkin feature to evaluate the application software. An alternative to Gherkin is using the **pytest** tool with the **pytest-bdd** plug-in to run the acceptance tests.

The two big advantages of Gherkin are the ability to structure the feature descriptions into scenarios and write the descriptions in English (or any other natural language). Framing the expected behavior into discrete operating scenarios forces us to think clearly about how the application or module is used. Writing in English (or other natural languages) makes it easier to share definitions with other people to confirm our understanding. It also helps to keep the definition of done focused on the problem domain without devolving into technical considerations and programming.

Each scenario can have three steps: Given, When, and Then. The Given step defines a context. The When step defines an action or a request of the software. The Then step defines the expected results. These step definitions can be as complex as needed, often involving multiple clauses joined with And. Examples can be provided in tables to avoid copying and pasting a scenario with a different set of values. A separate module provides Python implementations for the English-language step text.

See `https://behave.readthedocs.io/en/stable/gherkin.html#gherkin-feature-tes` `ting-language` for numerous examples of scenarios written in Gherkin.

Start this part of elaboration by creating a `tests/features/project.feature` file based on the overview description. Don't use a boring name like `project`. A complex project may have multiple features, so the feature file names should reflect the features.

To use **pytest**, write one (or more) acceptance test scripts in the `tests` directory.

The features are supported by **steps**. These steps are in modules in the `tests/steps` directory. A `tests/steps/hw_cli.py` module provides the necessary Python definitions for the steps in the feature file. The names of the modules don't matter; we suggest something like `hw_cli` because it implements the steps for a hello-world command-line interface.

The underlying mechanism is used by the **Behave** tool are function decorators. These match text from the feature file to define the function that implements that step. These can have wildcard-matching to permit flexibility in wording. The decorator can also parse out parameter values from the text.

A `tests/environment.py` file is required, but it can be empty for simple tests. This file provides a testing context, and is where some functions used by the **Behave** tool to control test setup and teardown are defined.

As soon as scenarios have been written, it makes sense to run the **Behave** tool to see the acceptance test fail. Initially, this lets you debug the step definitions.

For this application, the steps must properly execute the application program and capture the output file. Because the application doesn't exist yet, a test failure at this point is expected.

The feature files with the application scenarios are a working definition of done. When the test suite runs, it will show whether or not the software works. Starting with features that fail to work means the rest of the construction phase will be debugging the failures and fixing the software until the application passes the acceptance test suite.

In *Project 0 – Hello World with test cases* we'll look at an example of a Gherkin-language feature, the matching step definitions, and a `tox.ini` to run the test suite.

Elaboration, part 2: define components and tests

The acceptance test suite is often relatively "coarse" – the tests exercise the application as a whole, and avoid internal error conditions or subtle edge cases. The acceptance test suite rarely exercises all of the individual software components. Because of this, it can be difficult to debug problems in complex applications without detailed unit tests for each unit — each package, module, class, and function.

After writing the general acceptance test suite, it helps to do two things. First, start writing some skeleton code that's likely to solve the problem. The class or function will contain a docstring explaining the idea. Optionally, it can have a body of the `pass` statement. After writing this skeleton, the second step is to expand on the docstring ideas by writing unit tests for the components.

Let's assume we've written a scenario with a step that will execute an application named `src/hello_world.py`. We can create this file and include a skeleton class definition like this:

```
class Greeting:
    """
```

```
    Created with a greeting text.
    Writes the text to stdout.

    ..  todo:: Finish this
    """
    pass
```

This example shows a class with a design idea. This needs to be expanded with a clear statement of expected behaviors. Those expectations should take the form of unit tests for this class.

Once some skeletons and tests are written, the **pytest** tool can be used to execute those tests.

The unit tests will likely fail because the skeleton code is incomplete or doesn't work. In the cases where tests are complete, but classes don't work, you're ready to start the construction phase.

In the cases where the design isn't complete, or the tests are fragmentary, it makes sense to remain in the elaboration phase for those classes, modules, or functions. Once the tests are understood, construction has a clear and achievable goal.

We don't always get the test cases right the first time, we must change them as we learn. We rarely get the working code right the first time. If the test cases come first, they make sure we have a clear goal.

In some cases, the design may not be easy to articulate without first writing some "spike solution" to explore an alternative. Once the spike works, it makes sense to write tests to demonstrate the code works.

See http://www.extremeprogramming.org/rules/spike.html for more on creating spike solutions.

At this point, you have an idea of how the software will be designed. The test cases are a way to formalize the design into a goal. It's time to begin construction.

Construction

The construction phase finishes the class and function (and module and package) definitions started in the elaboration phase. In some cases, test cases will need to be added as the definitions expand.

As we get closer to solving the problem, the number of tests passed will grow.

The number of tests may also grow. It's common to realize the sketch of a class definition is incomplete and requires additional classes to implement the **State** or **Strategy** design pattern. As another example, we may realize subclasses are required to handle special cases. This new understanding will change the test suite.

When we look at our progress over several days, we should see that the number of tests pass approaches the total number of tests.

How many tests do we need? There are strong opinions here. For the purposes of showing high-quality work, tests that exercise 100% of the code are a good starting point. For some industries, a more strict rule is to cover 100% of the logic paths through the code. This higher standard is often used for applications like robotics and health care where the consequences of a software failure may involve injury or death.

Transition

For enterprise applications, there is a transition from the development team to formal operations. This usually means a deployment into a production environment with the real user community and their data.

In organizations with good Continuous Integration/Continuous Deployment (CI/CD) practices, there will be a formalized execution of the `tox` command to make sure everything works: all the tests pass.

In some enterprises, the `make html` command will also be run to create the documentation.

Often, the technical operations team will need specific topics in the documentation and the `README.md` file. Operations staff may have to diagnose and troubleshoot problems with hundreds of applications, and they will need very specific advice in places where they can find it immediately. We won't emphasize this in this book, but as we complete our projects, it's important to think that our colleagues will be using this software, and we want their work life to be pleasant and productive.

The final step is to post your project to your public repository of choice.

You have completed part of your portfolio. You'll want potential business partners or hiring managers or investors to see this and recognize your level of skill.

We can view a project as a sequence of steps. We can also view a project as a deliverable set of files created by those steps. In the next section, we'll look over the deliverables in a little more detail.

List of deliverables

We'll take another look at the project, this time from the view of what files will be created. This will parallel the outline of the activities shown in the previous section.

The following outline shows many of the files in a completed project:

- The documentation in the docs directory. There will be other files in there, but you'll be focused on the following files:

    - The Sphinx index.rst starter file with references to overview and API sections.

    - An overview.rst section with a summary of the project.

    - An api.rst section with .. automodule:: commands to pull in documentation from the application.

- A set of test cases in the tests directory.

    - Acceptance tests aimed at Behave (or the **pytest-bdd** plug-in for Gherkin). When using Behave, there will be two sub-directories: a features directory and a steps directory. Additionally, there will be an environment.py file.

    - Unit test modules written with the **pytest** framework. These all have a name that starts with `test_` to make them easy for **pytest** to find. Ideally, the **Coverage** tool is used to assure 100% of the code is exercised.

- The final code in the `src` directory. For some of the projects, a single module will be sufficient. Other projects will involve a few modules. (Developers familiar with Java or C++ often create too many modules here. The Python concept of *module* is more akin to the Java concept of *package*. It's not common Python practice to put each class definition into a separate module file.)

- Any JupyterLab notebooks can be in the `notebooks` folder. Not all projects use JupyterLab notebooks, so this folder can be omitted if there are no notebooks.

- A few other project files are in the top-level directory.

    - A `tox.ini` file should be used to run the **pytest** and **behave** test suites.

    - The `pyproject.toml` provides a number of pieces of information about the project. This includes a detailed list of packages and version numbers to be installed to run the project, as well as the packages required for development and testing. With this in place, the **tox** tool can then build virtual environments using the `requirements.txt` or the **pip-tools** tool to test the project. As a practical matter, this will also be used by other developers to create their working desktop environment.

    - An `environment.yml` can help other developers use **conda** to create their environment. This will repeat the contents of `requirements-dev.txt`. For a small team, it isn't helpful. In larger enterprise work groups, however, this can help others join your project.

    - Also, a `README.md` (or `README.rst`) with a summary is essential. In many cases, this is the first thing people look at; it needs to provide an "elevator pitch" for the project (see `https://www.atlassian.com/team-playbook/plays/elevator-pitch`).

See `https://github.com/cmawer/reproducible-model` for additional advice on structuring complex projects.

We've presented the files in this order to encourage following an approach of writing documentation first. This is followed by creating test cases to assure the documentation will be satisfied by the programming.

We've looked at the development activities and a review of the products to be created. In the next section, we'll look at some suggested development tools.

Development tool installation

Many of the projects in this book are focused on data analysis. The tooling for data analysis is often easiest to install with the **conda** tool. This isn't a requirement, and readers familiar with the **PIP** tool will often be able to build their working environments without the help of the **conda** tool.

We suggest the following tools:

- **Conda** for installing and configuring each project's unique virtual environment.

- **Sphinx** for writing documentation.

- **Behave** for acceptance tests.

- **Pytest** for unit tests. The **pytest-cov** plug-in can help to compute test coverage.

- **Pip-Tool** for building a few working files from the pyproject.toml project definition.

- **Tox** for running the suite of tests.

- **Mypy** for static analysis of the type annotations.

- **Flake8** for static analysis of code, in general, to make sure it follows a consistent style.

One of the deliverables is the pyproject.toml file. This has all of the metadata about the project in a single place. It lists packages required by the application, as well as the tools used for development and testing. It helps to pin exact version numbers, making it easier for someone to rebuild the virtual environment.

Some Python tools — like PIP — work with files derived from the `pyproject.toml` file. The **pip-tools** creates these derived files from the source information in the TOML file.

For example, we might use the following output to extract the development tools information from `pyproject.toml` and write it to `requirements-dev.txt`.

```
% conda install -c conda-forge pip-tools
% pip-compile --extra=dev --output-file=requirements-dev.txt
```

It's common practice to then use the `requirements-dev.txt` to install packages like this:

```
% conda install --file requirements-dev.txt --channel=conda-forge
```

This will try to install all of the named packages, pulled from the community `conda-forge` channel.

Another alternative is to use PIP like this:

```
% python -m pip install --r requirements-dev.txt
```

This environment preparation is an essential ingredient in each project's inception phase. This means the `pyproject.toml` is often the first deliverable created. From this, the `requirements-dev.txt` is extracted to build environments.

To make the preceding steps and deliverables more specific, we'll walk through an initial project. This project will help show how the remaining projects should be completed.

Project 0 – Hello World with test cases

This is our first project. This project will demonstrate the pattern for all of the book's projects. It will include these three elements.

- **Description**: The description section will set out a problem, and why a user needs software to solve it. In some projects, the description will have very specific details. Other projects will require more imagination to create a solution.

- **Approach**: The approach section will offer some guidance on architectural and design choices. For some projects there are trade-offs, and an **Extras** section will explore some of the other choices.

- **Deliverables**: The deliverables section lists the expectations for the final application or module. It will often provide a few Gherkin feature definitions.

For this initial project, the description isn't going to be very complicated. Similarly, the approach part of this first project will be brief. We'll dwell on the deliverables with some additional technical discussion.

Description

The problem the users need to solve is how best to bring new developers on board. A good onboarding process helps our users by making new members of the team as productive as quickly as possible. Additionally, a project like this can be used for experienced members to introduce them to new tools.

We need to guide our team members in installing the core set of development tools, creating a working module, and then displaying their completed work at the end of a sprint. This first project will use the most important tools and assure that everyone has a common understanding of the tools and the deliverables.

Each developer will build a project to create a small application. This application will have a command-line interface (CLI) to write a cheerful greeting.

The expectations are shown in the following example:

```
% python src/hello_world.py --who "World"
Hello, World!
```

This example shows how running the application with a command-line parameter of `--who "world"` produces a response on the console.

Approach

For this project, the objective is to create a Python application module. The module will need several internal functions. The functions can be combined into a class, if that seems more appropriate. The functions are these:

- A function to parse the command-line options. This will use the `argparse` module. The default command-line argument values are available in `sys.argv`.

- A function to write a cheerful greeting. This is, perhaps, only a single line of code.

- An overall function with an obvious name like `main()` to get the options and write the greeting.

The module, as a whole, will have the function (or class) definitions. It will also have an `if __name__ == "__main__":` block. This block will guard the evaluation of the expression `main()` to make the module easier to unit test.

 This is quite a bit of engineering for a simple problem. Some might call it over-engineering. The idea is to create something with enough complexity that more than one unit test case is required.

Deliverables

As noted above in *List of deliverables*, there are a number of deliverable files for projects in general. Here are the suggested files for this project:

- `README.md` summarizes the project.

- `pyproject.toml` defines the project, including development tools, test tools, and other dependencies.

- `docs` contains the documentation. As described above, this should be built by the `sphinx-quickstart` tool and should contain at least an overview and an API section.

- `tests` contains test cases; the files include the following:

  - `test_hw.py` contains unit tests for the module's functions or classes.

- features/hello_world.feature has an overall acceptance test as a collection of scenarios.

- steps/hw_cli.py has Python definitions for the steps in the scenarios.

- environment.py contains functions to control **behave**'s test setup and teardown. For simple projects, it may be empty.

- tox.ini configuration for the **tox** tool to run the complete test suite.

- src contains the hello_world.py module.

We'll look at a few of these files in detail in the following sub-sections.

The pyproject.toml project file

The pyproject.toml file contains a great deal of project metadata in a single location. The minimal content of this file is a description of the "build-system" used to build and install the package.

For the purposes of this book, we can use the following two lines to define the build system:

```
[build-system]
requires = ["setuptools", "wheel"]  # PEP 508 specifications.
```

This specifies the use of the setuptools module to create a "wheel" with the project's code. The pyproject.toml doesn't need to define the distribution package in any more detail. This book doesn't emphasize the creation of a distribution package or the management of packages with the Python Package Index, PyPI.

The rest of this file should have information about the project. You can include a section like the following:

```
[project]
name = "project_0"
version = "1.0.0"
authors = [
```

```
    {name = "Author", email = "author@email.com"},
]
description = "Real-World Python Projects -- Project 0."
readme = "README.md"
requires-python = ">=3.10"
```

Clearly, you'll want to update the `authors` section with your information. You may be using a newer version of Python and may need to change the `requires-python` string to specify the minimum version required for your unique solution.

The `[project]` section needs three other pieces of information:

- The packages required to execute your application.

- Any packages or tools required to test your application.

- Any packages or tools required to develop your application.

These three dependencies are organized as follows:

```
dependencies = [
    # Packages required -- None for Project Zero.
]
[project.optional-dependencies]
dev = [
    # Development tools to work on this project
    "sphinx==7.0.1",
    "sphinxcontrib-plantuml==0.25",
    "pip-tools==6.13.0"
]
test = [
    # Testing tools to test this project
    "pytest==7.2.0",
    "tox==4.0.8",
```

```
    "behave==1.2.6"
]
```

The dependencies line lists the dependencies required to execute the application. Some projects — like this one — rely on the standard library, and nothing more needs to be added. The [project.optional-dependencies] section contains two lists of additional packages: those required for development, and those required for testing.

Note that we've put specific version numbers in this file so that we can be absolutely certain what packages will be used. As these packages evolve, we'll need to test newer versions and upgrade the dependencies.

If you see the version numbers in this book are behind the current state of the art on PyPI or Conda-Forge, feel free to use up-to-date versions.

It helps to use the **pip-compile** command. This is installed as part of **pip-tools**. This command create extract files from the pyproject.toml file for use by **pip** or **conda**.

For developers, we often want to install all of the "extras." This usually means executing the following command to create a requirements-dev.txt file that can be used to build a development environment.

```
% pip-compile --all-extras -o requirements-dev.txt
```

In order to run the **tox** tool, it's common to also create a testing-only subset of the required packages and tools. Use the following command:

```
% pip-compile --extra test -o requirements.txt
```

This creates the requirements.txt to be used to detect manage virtual environments used by **tox** for testing.

The docs directory

As noted above in *Suggested project sprints* this directory should be built with `sphinx-quickstart`. After the initial set of files is created, make the following changes:

- Add a `api.rst` file as a placeholder for the Sphinx-generated API documentation. This will use the `.. automodule::` directive to extract documentation from your application.

- Add a `overview.rst` file with an overview of the project.

- Update the `index.rst` to include these two new files in the table of contents.

- Update the `conf.py` to append the `src` directory to `sys.path`. Also, the `sphinx.ext.autodoc` extension needs to be added to the `extensions` setting in this file.

The `make html` command in the `docs` directory can be used to build the documentation.

The tests/features/hello_world.feature file

The `features` directory will have Gherkin-language definitions of the features. Each feature file will contain one or more scenarios. For larger projects, these files often start with statements lifted from problem descriptions or architectural overviews that are later refined into more detailed steps to describe an application's behavior.

For this project, one of the feature files should be `features/hello_world.feature`. The contents of this file should include a description of the feature and at least one scenario. It would look like the following example:

```
Feature: The Cheerful Greeting CLI interface provides a greeting
    to a specific name.


Scenario: When requested, the application writes the greeting message.
  When we run command "python src/hello_world.py"
  Then output has "Hello, World!"
```

There's no Given step in this scenario; there's no initialization or preparation required. Each of the steps has only a single clause, so there are no And steps, either.

 This example doesn't precisely match the example in the description. There are two possible reasons for this: one of the two examples is wrong, or, more charitably, this example hints at a second feature.

The idea implied by this example is there's a default behavior when no --who command-line option is provided. This suggests that a second scenario — one with the --who option should be added for this feature.

The tests/steps/hw_cli.py module

The steps directory contains modules that define the natural-language phrases in the feature files. In the hello_world.feature file the When and Then steps had phrases written out in plain English:

- We run the command "python src/hello_world.py"

- Output has "Hello, World!"

The steps/hw_cli.py module will map the step's phrases to Python functions. It works by using decorators and pattern-matching to specify the type of step (@given, @when, or @then) and the text to match. The presence of {parameter} in the text will match the text and provide the value matched to the step function as an argument. The function names are irrelevant and are often step_impl().

Generally, the @given steps will accumulate parameter values in the test context object. Best practices suggest there should be only one @when step; this will perform the required operation. For this project, it will run the application and gather the output files. The @then steps can use assert statements to compare actual results against the expected results shown in the feature file.

Here's how the steps/hw_cli.py module might look:

```
import subprocess
```

```python
import shlex
from pathlib import Path


@when(u'we run command "{command}"')
def step_impl(context, command):
    output_path = Path("output.log")
    with output_path.open('w') as target:
        status = subprocess.run(
            shlex.split(command),
            check=True, text=True, stdout=target, stderr=subprocess.STDOUT)
    context.status = status
    context.output = output_path.read_text()
    output_path.unlink()


@then(u'output has "{expected_output}"')
def step_impl(context, expected_output):
    assert context.status.returncode == 0
    assert expected_output in context.output
```

This assumes a relatively small output file that can be collected in memory. For a larger file, it would make sense for the @when step to create a temporary file and save the file object in the context. The @then step can read and close this file. The `tempfile` module is handy for creating files that will be deleted when closed.

An alternative is to create a `Path` object and save this object in the context. The @when step can write output to this path. The @then step can read and examine the contents of the file named by the `Path` object.

When a test step detects a problem with an `assert` statement, it may not finish completely. The approach of using a `Path` object requires some care to be sure the temporary files are deleted. The `environment.py` module can define an `after_scenario(context, scenario)` function to remove temporary files.

The tests/environment.py file

This module will contain some function definitions used by **behave**. For this project, it will be empty. The module must be present; a module docstring is appropriate to explain that it's empty.

The tests/steps module for this example will have examples that can be refactored into two potentially reusable functions for executing an application and checking the output from an application for specific text. This additional design effort isn't part of this project. You may find it helpful to do this refactoring after completing several of these projects.

Once the features, steps, and environment are in place, the **behave** program can be used to test the application. If there's no application module, the tests will fail. Creating a skeleton application module in the src directory will allow the test case to execute and fail because the output isn't what was expected.

The tests/test_hw.py unit tests

A unit test can be implemented as a **pytest** function that uses a fixture, capsys, to capture the system output. The unit test case expects the application to have a main() function that parses the command-line options.

Here's one suggested unit test function:

```python
import hello_world


def test_hw(capsys):
    hello_world.main([])
    out, err = capsys.readouterr()
    assert "Hello, World!" in out
```

Note the test for the main() function provides an explicit empty list of argument values. It is essential to override any value for sys.argv that might be present when **pytest** is running.

The hello_world module is imported by this test. There are two important consequences of this import:

- The `src/hello_world.py` module must have an `if __name__ == "__main__":` section. A simple Python script (without this section) will execute completely when imported. This can make testing difficult.

- The `src` directory must be part of the `PYTHONPATH` environment variable. This is handled by the `tox.ini` file.

This test will tolerate additional output in addition to the required cheerful greeting. It might make sense to use something like `"Hello, World!" == out.strip()`.

The implementation details of the `main()` function are opaque to this test. This `main()` function could create an instance of a class; it could use a static method of a class, also.

The src/tox.ini file

Now that the tests exist, we can run them. The **tox** (and **nox**) tools are ideal for running a suite of tests.

Here's an example `tox.ini` file:

```
[tox]
min_version = 4.0
skipsdist = true


[testenv]
deps = pip-tools
    pytest
    behave
commands_pre = pip-sync requirements.txt
setenv =
    PYTHONPATH=src
commands =
    pytest tests
```

```
behave tests
```

This file lists the tools used for testing: **pip-tools**, **pytest**, and **behave**. It provides the setting for the PYTHONPATH. The commands_pre will prepare the the virtual environment using the **pip-sync** command that is part of the **pip-tools** package. The given sequence of commands defines the test suite.

The src/hello_world.py file

This is the desired application module. The test framework is helpful to confirm that it really does work, and — more importantly — it meets the definition of done provided in the *.feature files.

As we noted above, the unit tests will import this app as a module. The acceptance test, in contrast, will run the app. This means the if __name__ == "__main__": section is essential.

For a small application like this, the real work of the application should be encapsulated in a main() function. This allows the main module to end with the following snippet:

```
if __name__ == "__main__":
    main()
```

This assures that the module will not take off and start running when imported. It will only do useful work when invoked from the command line.

Definition of done

This project is tested by running the tox command.

When all of the tests execute, the output will look like this:

```
(projectbook) slott@MacBookPro-SLott project_0 % tox
py: commands[0]> pytest tests
...
py: commands[1]> behave tests
...
```

```
py: OK (0.96=setup[0.13]+cmd[0.53,0.30] seconds)
congratulations :) (1.55 seconds)
```

This output has elided the details from **pytest** and **behave**. The output from the **tox** tool is the important summary py: OK. This tells us all the tests passed.

Once this is complete, we can run the following to create the API documentation:

```
% (cd docs; make html)
```

It can help to wrap the two commands with () so the cd docs command doesn't leave the console session in the docs directory. Some developers prefer to have two windows open: one in the top-level directory to run the **tox** tool and one in the docs subdirectory to run the **make** commands for the **sphinx** tool.

Summary

In this chapter, we've looked at the following topics:

- An overview of the software quality principles that we'll try to emphasize.

- A suggested approach to completing the project as a sequence of project sprints.

- A general overview of the list of deliverables for each project.

- The tools suggested for creating these examples.

- A sample project to act as a template for subsequent projects.

After creating this initial project, the next chapter will look at the general collection of projects. The idea is to create a complete data analysis tool set with a number of closely-related projects.

Extras

Here are some ideas for you to add to this project.

Static analysis - mypy, flake8

There are several common static analysis tools that are as essential as automated testing:

- **mypy** checks type annotations to be sure the functions and classes will interact properly.

- **flake8** does other syntax checks to make sure the code avoids some of the more common Python mistakes.

- **black** can be used to check the formatting to make sure it follows the recommended style. The `black` application can also be used to reformat a new file.

- **isort** can be used to put a long collection of `import` statements into a consistent order.

Once the application passes the functional tests in the `*.feature` files, these additional non-functional tests can be applied. These additional tests are often helpful for spotting more nuanced problems that can make a program difficult to adapt or maintain.

CLI features

The command-language interface permits a single option, the `--who` option, to provide a name.

It makes sense to add a scenario to exercise this option.

What should happen with the `--who` is provided without a value? Is the following appropriate?

```
(projectbook) slott@MacBookPro-SLott project_0 % python src/hello_world.py
    --who
```

```
usage: hello_world.py [-h] [--who WHO]
hello_world.py: error: argument --who/-w: expected one argument
```

Should the help be extended to clarify what's required?

Consider adding the following scenarios (and the implementing code):

- Add a scenario for the `--help` option, which is provided automatically by the `argparse` module.

- Add a scenario for the `--who` with no value error.

Logging

Consider a more complex application where additional debugging output might be helpful. For this, it's common to add a `--verbose` option to set the logging level to `logging.DEBUG` instead of a default level of `logging.INFO`.

Adding this option requires adding logging capabilities. Consider making the following changes to this module:

- Import the `logging` module and create a global logger for the application.

- Update the `main()` function to set the logger's level based on the options.

- Update the `__name__ == "__main__":` block to have two lines: `logging.basicConfig()` and `main()`. It's best to keep logging configuration isolated from the rest of the application processing.

Cookiecutter

The `cookiecutter` project (see `https://cookiecutter.readthedocs.io/en/stable/`) is a way to build a template project. This can help team members get started by sharing a single template. As tool versions or solution architectures change, additional cookie-cutter templates can be developed and used.

There are thousands of cookie-cutter templates available. It can be difficult to locate one that's suitably simple. It may be better to create your own and add to it as new concepts are introduced in later chapters.

2
Overview of the Projects

Our general plan is to craft analytic, decision support modules and applications. These applications support decision-making by providing summaries of available data to the stakeholders. Decision-making spans a spectrum from uncovering new relationships among variables to confirming that data variation is random noise within narrow limits. The processing will start with acquiring data and moving it through several stages until statistical summaries can be presented.

The processing will be decomposed into several stages. Each stage will be built as a core concept application. There will be subsequent projects to add features to the core application. In some cases, a number of features will be added to several projects all combined into a single chapter.

The stages are inspired by the **Extract-Transform-Load** (ETL) architectural pattern. The design in this book expands on the ETL design with a number of additional steps. The words have been changed because the legacy terminology can be misleading. These features — often required for real-world pragmatic applications — will be inserted as additional stages in a pipeline.

Once the data is cleaned and standardized, then the book will describe some simple statistical models. The analysis will stop there. You are urged to move to more advanced books that cover AI and machine learning.

There are 22 distinct projects, many of which build on previous results. It's not required to do all of the projects in order. When skipping a project, however, it's important to read the description and deliverables for the project being skipped. This can help to more fully understand the context for the later projects.

This chapter will cover our overall architectural approach to creating a complete sequence of data analysis programs. We'll use the following multi-stage approach:

- Data acquisition

- Inspection of data

- Cleaning data; this includes validating, converting, standardizing, and saving intermediate results

- Summarizing, and modeling data

- Creating more sophisticated statistical models

The stages fit together as shown in *Figure 2.1*.

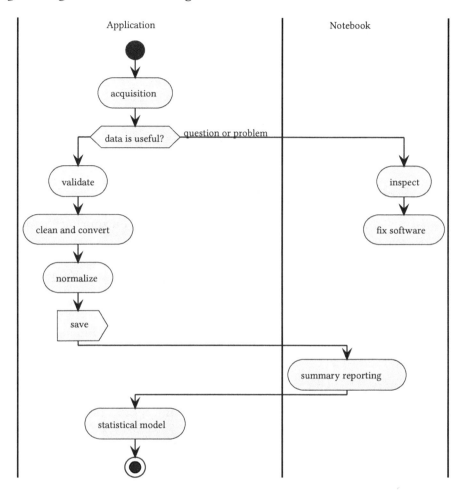

Figure 2.1: Data Analysis Pipeline

A central idea behind this is *separation of concerns*. Each stage is a distinct operation, and each stage may evolve independently of the others. For example, there may be multiple sources of data, leading to several distinct data acquisition implementations, each of which creates a common internal representation to permit a single, uniform inspection tool.

Similarly, data cleansing problems seem to arise almost randomly in organizations, leading to a need to add distinct validation and standardization operations. The idea is to allocate responsibility for semantic special cases and exceptions in this stage of the pipeline.

One of the architectural ideas is to mix automated applications and a few manual JupyterLab notebooks into an integrated whole. The notebooks are essential for troubleshooting questions or problems. For elegant reports and presentations, notebooks are also very useful. While Python applications can produce tidy PDF files with polished reporting, it seems a bit easier to publish a notebook with analysis and findings.

We'll start with the acquisition stage of processing.

General data acquisition

All data analysis processing starts with the essential step of acquiring the data from a source.

The above statement seems almost silly, but failures in this effort often lead to complicated rework later. It's essential to recognize that data exists in these two essential forms:

- Python objects, usable in analytic programs. While the obvious candidates are numbers and strings, this includes using packages like **Pillow** to operate on images as Python objects. A package like **librosa** can create objects representing audio data.

- A serialization of a Python object. There are many choices here:

  - Text. Some kind of string. There are numerous syntax variants, including CSV, JSON, TOML, YAML, HTML, XML, etc.

  - Pickled Python Objects. These are created by the `pickle` module.

  - Binary Formats. Tools like Protobuf can serialize native Python objects into a stream of bytes. Some YAML extensions, similarly, can serialize an object in a binary format that isn't text. Images and audio samples are often stored in compressed binary formats.

The format for the source data is — almost universally — not fixed by any rules or conventions. Writing an application based on the assumption that source data is **always** a CSV-format file can lead to problems when a new format is required.

It's best to treat all input formats as subject to change. The data — once acquired — can be saved in a common format used by the analysis pipeline, and independent of the source format (we'll get to the persistence in *Clean, validate, standardize, and persist*).

We'll start with Project 1.1: "Acquire Data". This will build the Data Acquisition Base Application. It will acquire CSV-format data and serve as the basis for adding formats in later projects.

There are a number of variants on how data is acquired. In the next few chapters, we'll look at some alternative data extraction approaches.

Acquisition via Extract

Since data formats are in a constant state of flux, it's helpful to understand how to add and modify data formats. These projects will all build on Project 1.1 by adding features to the base application. The following projects are designed around alternative sources for data:

- Project 1.2: "Acquire Web Data from an API". This project will acquire data from web services using JSON format.

- Project 1.3: "Acquire Web Data from HTML". This project will acquire data from a web page by scraping the HTML.

- Two separate projects are part of gathering data from a SQL database:

  - Project 1.4: "Build a Local Database". This is a necessary sidebar project to build a local SQL database. This is necessary because SQL databases accessible by the public are a rarity. It's more secure to build our own demonstration database.

  - Project 1.5: "Acquire Data from a Local Database". Once a database is available, we can acquire data from a SQL extract.

These projects will focus on data represented as text. For CSV files, the data is text; an application must convert it to a more useful Python type. HTML pages, also, are pure text. Sometimes, additional attributes are provided to suggest the text should be treated as a number. A SQL database is often populated with non-text data. To be consistent, the SQL data should be serialized as text. The acquisition applications all share a common approach of working with text.

These applications will also minimize the transformations applied to the source data. To process the data consistently, it's helpful to make a shift to a common format. As we'll see in *Chapter 3, Project 1.1: Data Acquisition Base Application* the NDJSON format provides a useful structure that can often be mapped back to source files.

After acquiring new data, it's prudent to do a manual inspection. This is often done a few times at the start of application development. After that, inspection is only done to diagnose problems with the source data. The next few chapters will cover projects to inspect data.

Inspection

Data inspection needs to be done when starting development. It's essential to survey new data to be sure it really is what's needed to solve the user's problems. A common frustration is incomplete or inconsistent data, and these problems need to be exposed as soon as possible to avoid wasting time and effort creating software to process data that doesn't really exist.

Additionally, data is inspected manually to uncover problems. It's important to recognize that data sources are in a constant state of flux. As applications evolve and mature, the data provided for analysis will change. In many cases, data analytics applications discover other enterprise changes after the fact via invalid data. It's important to understand the evolution via good data inspection tools.

Inspection is an inherently manual process. Therefore, we're going to use JupyterLab to create notebooks to look at the data and determine some basic features.

In rare cases where privacy is important, developers may not be allowed to do data inspection. More privileged people — with permission to see payment card or healthcare details — may be part of data inspection. This means an inspection notebook may be something created by a developer for use by stakeholders.

In many cases, a data inspection notebook can be the start of a fully-automated data cleansing application. A developer can extract notebook cells as functions, building a module that's usable from both notebook and application. The cell results can be used to create unit test cases.

The stage in the pipeline requires a number of inspection projects:

- Project 2.1: "Inspect Data". This will build a core data inspection notebook with enough features to confirm that some of the acquired data is likely to be valid.

- Project 2.2: "Inspect Data: Cardinal Domains". This project will add analysis features for measurements, dates, and times. These are cardinal domains that reflect measures and counts.

- Project 2.3: "Inspect Data: Nominal and Ordinary Domains". This project will add analysis features for text or coded numeric data. This includes nominal data and ordinal numeric domains. It's important to recognize that US Zip Codes are digit strings, not numbers.

- Project 2.4: "Inspect Data: Reference Data". This notebook will include features to find reference domains when working with data that has been normalized and decomposed into subsets with references via coded "key" values.

- Project 2.5: "Define a Reusable Schema". As a final step, it can help define a formal schema, and related metadata, using the JSON Schema standard.

While some of these projects seem to be one-time efforts, they often need to be written with some care. In many cases, a notebook will need to be reused when there's a problem. It helps to provide adequate explanations and test cases to help refresh someone's memory on details of the data and what are known problem areas. Additionally, notebooks may serve as examples for test cases and the design of Python classes or functions to automate cleaning, validating, or standardizing data.

After a detailed inspection, we can then build applications to automate cleaning, validating, and normalizing the values. The next batch of projects will address this stage of the pipeline.

Clean, validate, standardize, and persist

Once the data is understood in a general sense, it makes sense to write applications to clean up any serialization problems, and perform more formal tests to be sure the data really is valid. One frustratingly common problem is receiving duplicate files of data; this can happen when scheduled processing was disrupted somewhere else in the enterprise, and a previous period's files were reused for analysis.

The validation testing is sometimes part of cleaning. If the data contains any unexpected invalid values, it may be necessary to reject it. In other cases, known problems can be resolved as part of analytics by replacing invalid data with valid data. An example of this is US Postal Codes, which are (sometimes) translated into numbers, and the leading zeros are lost.

These stages in the data analysis pipeline are described by a number of projects:

- Project 3.1: "Clean Data". This builds the data cleaning base application. The design details can come from the data inspection notebooks.

- Project 3.2: "Clean and Validate". These features will validate and convert numeric fields.

- Project 3.3: "Clean and Validate Text and Codes". The validation of text fields and numeric coded fields requires somewhat more complex designs.

- Project 3.4: "Clean and Validate References". When data arrives from separate sources, it is essential to validate references among those sources.

- Project 3.5: "Standardize Data". Some data sources require standardizing to create common codes and ranges.

- Project 3.6: "Acquire and Clean Pipeline". It's often helpful to integrate the acquisition, cleaning, validating, and standardizing into a single pipeline.

- Project 3.7: "Acquire, Clean, and Save". One key architectural feature of this pipeline is saving intermediate files in a common format, distinct from the data sources.

- Project 3.8: "Data Provider Web Service". In many enterprises, an internal web service and API are expected as sources for analytic data. This project will wrap the data acquisition pipeline into a RESTful web service.

In these projects, we'll transform the text values from the acquisition applications into more useful Python objects like integers, floating-point values, decimal values, and date-time values.

Once the data is cleaned and validated, the exploration can continue. The first step is to summarize the data, again, using a Jupyter notebook to create readable, publishable reports and presentations. The next chapters will explore the work of summarizing data.

Summarize and analyze

Summarizing data in a useful form is more art than technology. It can be difficult to know how best to present information to people to help them make more valuable, or helpful decisions.

There are a few projects to capture the essence of summaries and initial analysis:

- Project 4.1: "A Data Dashboard". This notebook will show a number of visual analysis techniques.

- Project 4.2: "A Published Report". A notebook can be saved as a PDF file, creating a report that's easily shared.

The initial work of summarizing and creating shared, published reports sets the stage for more formal, automated reporting. The next set of projects builds modules that provide deeper and more sophisticated statistical models.

Statistical modeling

The point of data analysis is to digest raw data and present information to people to support their decision-making. The previous stages of the pipeline have prepared two important things:

- Raw data has been cleaned and standardized to provide data that are relatively easy to analyze.

- The process of inspecting and summarizing the data has helped analysts, developers, and, ultimately, users understand what the information means.

The confluence of data and deeper meaning creates significant value for an enterprise. The analysis process can continue as more formalized statistical modeling. This, in turn, may lead to artificial intelligence (AI) and machine learning (ML) applications.

The processing pipeline includes these projects to gather summaries of individual variables as well as combinations of variables:

- Project 5.1: "Statistical Model: Core Processing". This project builds the base application for applying statistical models and saving parameters about the data. This will focus on summaries like mean, median, mode, and variance.

- Project 5.2: "Statistical Model: Relationships". It's common to want to know the relationships among variables. This includes measures like correlation among variables.

This sequence of stages produces high-quality data and provides ways to diagnose and debug problems with data sources. The sequence of projects will illustrate how automated solutions and interactive inspection can be used to create useful, timely, insightful reporting and analysis.

Data contracts

We will touch on data contracts at various stages in this pipeline. This application's data acquisition, for example, may have a formalized contract with a data provider. It's also possible that an informal data contract, in the form of a schema definition, or an API is all that's available.

In *Chapter 8, Project 2.5: Schema and Metadata* we'll consider some schema publication concerns. In *Chapter 11, Project 3.7: Interim Data Persistence* we'll consider the schema provided to downstream applications. These two topics are related to a formal data contract, but this book won't delve deeply into data contracts, how they're created, or how they might be used.

Summary

This data analysis pipeline moves data from sources through a series of stages to create clean, valid, standardized data. The general flow supports a variety of needs and permits a great deal of customization and extension.

For developers with an interest in data science or machine learning, these projects cover what is sometimes called the "data wrangling" part of data science or machine learning. It can be a significant complication as data is understood and differences among data sources are resolved and explored. These are the — sometimes difficult — preparatory steps prior to building a model that can be used for AI decision-making.

For readers with an interest in the web, this kind of data processing and extraction is part of presenting data via a web application API or website. Project 3.7 creates a web server, and will be of particular interest. Because the web service requires clean data, the preceding projects are helpful for creating data that can be published.

For folks with an automation or IoT interest, *Part 2* explains how to use Jupyter Notebooks to gather and inspect data. This is a common need, and the various steps to clean, validate, and standardize data become all the more important when dealing with real-world devices subject to the vagaries of temperature and voltage.

We've looked at the following multi-stage approach to doing data analysis:

- Data Acquisition

- Inspection of Data

- Clean, Validate, Standardize, and Persist

- Summarize and Analyze

- Create a Statistical Model

This pipeline follows the Extract-Transform-Load (ETL) concept. The terms have been changed because the legacy words are sometimes misleading. Our acquisition stage overlaps with what is understood as the "Extract" operation. For some developers, Extract is limited to database extracts; we'd like to go beyond that to include other data source transformations. Our cleaning, validating, and standardizing stages are usually combined into the "Transform" operation. Saving the clean data is generally the objective of "Load"; we're not emphasizing a database load, but instead, we'll use files.

Throughout the book, we'll describe each project's objective and provide the foundation of a sound technical approach. The details of the implementation are up to you. We'll enumerate the deliverables; this may repeat some of the information from *Chapter 1, Project Zero: A Template for Other Projects*. The book provides a great deal of information on acceptance test cases and unit test cases — the definition of done. By covering the approach, we've left room for you to design and implement the needed application software.

In the next chapter, we'll build the first data acquisition project. This will work with CSV-format files. Later projects will work with database extracts and web services.

3

Project 1.1: Data Acquisition Base Application

The beginning of the data pipeline is acquiring the raw data from various sources. This chapter has a single project to create a **command-line application** (**CLI**) that extracts relevant data from files in CSV format. This initial application will restructure the raw data into a more useful form. Later projects (starting in *Chapter 9, Project 3.1: Data Cleaning Base Application*) will add features for cleaning and validating the data.

This chapter's project covers the following essential skills:

- Application design in general. This includes an object-oriented design and the SOLID design principles, as well as functional design.

- A few CSV file processing techniques. This is a large subject area, and the project focuses on restructuring source data into a more usable form.

- CLI application construction.

- Creating acceptance tests using the Gherkin language and **behave** step definitions.

- Creating unit tests with mock objects.

We'll start with a description of the application, and then move on to talk about the architecture and construction. This will be followed by a detailed list of deliverables.

Description

Analysts and decision-makers need to acquire data for further analysis. In many cases, the data is available in CSV-formatted files. These files may be extracts from databases or downloads from web services.

For testing purposes, it's helpful to start with something relatively small. Some of the Kaggle data sets are very, very large, and require sophisticated application design. One of the most fun small data sets to work with is Anscombe's Quartet. This can serve as a test case to understand the issues and concerns in acquiring raw data.

We're interested in a few key features of an application to acquire data:

- When gathering data from multiple sources, it's imperative to convert it to a common format. Data sources vary, and will often change with software upgrades. The acquisition process needs to be flexible with respect to data sources and avoid assumptions about formats.

- A CLI application permits a variety of automation possibilities. For example, a CLI application can be "wrapped" to create a web service. It can be run from the command line manually, and it can be automated through enterprise job scheduling applications.

- The application must be extensible to reflect source changes. In many cases, enterprise changes are not communicated widely enough, and data analysis applications discover changes "the hard way" — a source of data suddenly includes unexpected or seemingly invalid values.

User experience

The **User Experience** (**UX**) will be a command-line application with options to fine-tune the data being gathered. This essential UX pattern will be used for many of this book's projects. It's flexible and can be made to run almost anywhere.

Our expected command line should look something like the following:

```
% python src/acquire.py -o quartet Anscombe_quartet_data.csv
```

The `-o quartet` argument specifies a directory into which the resulting extracts are written. The source file contains four separate series of data. Each of the series can be given an unimaginative name like `quartet/series_1.json`.

The positional argument, `Anscombe_quartet_data.csv`, is the name of the downloaded source file.

While there's only one file – at the present time – a good design will work with multiple input files and multiple source file formats.

In some cases, a more sophisticated "dashboard" or "control panel" application might be desirable as a way to oversee the operation of the data acquisition process. The use of a web-based API can provide a very rich interactive experience. An alternative is to use tools like **rich** or **Textual** to build a small text-based display. Either of these choices should be built as a wrapper that executes the essential CLI application as a subprocess.

Now that we've seen an overview of the application's purpose and UX, let's take a look at the source data.

About the source data

Here's the link to the dataset we'll be using:

```
https://www.kaggle.com/datasets/carlmcbrideellis/data-anscombes-quartet
```

You'll need to register with Kaggle to download this data.

The Kaggle URL presents a page with information about the CSV-formatted file. Clicking the **Download** button will download the small file of data to your local computer.

The data is available in this book's GitHub repository's data folder, also.

Once the data is downloaded, you can open the Anscombe_quartet_data.csv file to inspect the raw data.

The file contains four series of (x, y) pairs in each row. We can imagine each row as having $[(x_1, y_1), (x_2, y_2), (x_3, y_3), (x_4, y_4)]$. It is, however, compressed, as we'll see below.

We might depict the idea behind this data with an entity-relationship diagram as shown in *Figure 3.1.*

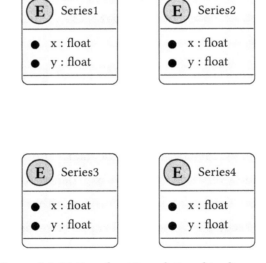

Figure 3.1: Notional entity-relationship diagram

Interestingly, the data is not organized as four separate (x, y) pairs. The downloaded file is organized as follows:

$$[x_{1,2,3}, y_1, y_2, y_3, x_4, y_4]$$

We can depict the actual source entity type in an ERD, as shown in *Figure 3.2*.

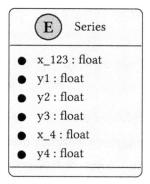

Figure 3.2: Source entity-relationship diagram

One part of this application's purpose is to disentangle the four series into separate files. This forces us to write some transformational processing to rearrange each row's data elements into four separate data sets.

The separate series can then be saved into four separate files. We'll look more deeply at the details of creating the separate files for a separate project in *Chapter 11, Project 3.7: Interim Data Persistence*. For this project, any file format for the four output files will do nicely; ND JSON serialization is often ideal.

We encourage you to take a look at the file before moving on to consider how it needs to be transformed into distinct output files.

Given this compressed file of source data, the next section will look at the expanded output files. These will separate each series to make them easier to work with.

About the output data

The ND JSON file format is described in `http://ndjson.org` and `https://jsonlines.org`. The idea is to put each individual entity into a JSON document written as a single physical line. This fits with the way the Python `json.dumps()` function works: if no value is provided for the `indent` parameter (or if the value is `indent=None`), the text will be as compact as possible.

The `series_1.json` output file should start like this:

```
{"x": "10.0", "y": "8.04"}
{"x": "8.0", "y": "6.95"}
{"x": "13.0", "y": "7.58"}
...
```

Each row is a distinct, small JSON document. The row is built from a subset of fields in the input file. The values are strings: we won't be attempting any conversions until the cleaning and validating projects in *Chapter 9, Project 3.1: Data Cleaning Base Application*.

We'll require the user who runs this application to create a directory for the output and provide the name of the directory on the command line. This means the application needs to present useful error messages if the directory doesn't actually exist. The `pathlib.Path` class is very helpful for confirming a directory exists.

Further, the application should be cautious about overwriting any existing files. The `pathlib.Path` class is very helpful for confirming a file already exists.

This section has looked at the input, processing, and output of this application. In the next section, we'll look at the overall architecture of the software.

Architectural approach

We'll take some guidance from the C4 model (`https://c4model.com`) when looking at our approach.

- **Context**: For this project, a context diagram would show a user extracting data from a source. You may find it helpful to draw this diagram.

- **Containers**: This project will run on the user's personal computer. As with the context, the diagram is small, but some readers may find it helpful to take the time to draw it.

- **Components**: We'll address these below.

- **Code**: We'll touch on this to provide some suggested directions.

We can decompose the software architecture into these two important components:

- `model`: This module has definitions of target objects. In this project, there's only a single class here.

- `extract`: This module will read the source document and creates model objects.

Additionally, there will need to be these additional functions:

- A function for parsing the command-line options.

- A `main()` function to parse options and do the file processing.

As suggested in *Chapter 1, Project Zero: A Template for Other Projects*, the initialization of logging will often look like the following example:

```
if __name__ == "__main__":
    logging.basicConfig(level=logging.INFO)
    main()
```

The idea is to write the `main()` function in a way that maximizes reuse. Avoiding logging initialization means other applications can more easily import this application's `main()` function to reuse the data acquisition features.

Initializing logging within the `main()` function can undo previous logging initialization. While there are ways to have a composite application tolerate each `main()` function doing yet another initialization of logging, it seems simpler to refactor this functionality outside the important processing.

For this project, we'll look at two general design approaches for the model and extract components. We'll utilize this opportunity to highlight the importance of adhering to SOLID design principles.

First, we'll show an object-oriented design using class definitions. After that, we'll show a functional design, using only functions and stateless objects.

Class design

One possible structure for the classes and functions of this application is shown in *Figure 3.3*.

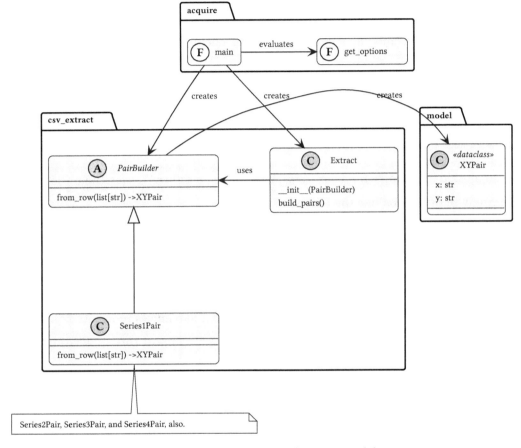

Figure 3.3: Acquisition Application Model

The model module contains a single class definition for the raw XYPair. Later, this is likely to expand and change. For now, it can seem like over-engineering.

The acquisition module contains a number of classes that collaborate to build XYPair objects for any of the four series. The abstract PairBuilder class defines the general features of creating an XYPair object.

Each subclass of the PairBuilder class has a slightly different implementation. Specifically, the `Series1Pair` class has a `from_row()` method that assembles a pair from the $x_{1,2,3}$ and y_1 values. Not shown in the diagram are the three other subclasses that use distinct pairs of columns to create `XYPair` objects from the four series.

> The diagram and most of the examples here use `list[str]` as the type for a row from a CSV reader.
>
> If a `csv.DictReader` is used, the source changes from `list[str]` to `dict[str, str]`. This small, but important, change will ripple throughout the examples.
>
> In many cases, it seems like a `csv.DictReader` is a better choice. Column names can be provided if the CSV file does not have names in the first row.
>
> We've left the revisions needed for this change as part of the design work for you.

The overall `Extract` class embodies the various algorithms for using an instance of the `PairBuilder` class and a row of source data to build `XYPair` instances. The `build_pair(list[str]) -> XYPair` method makes a single item from a row parsed from a CSV file.

The job of the `main()` function is to create instances of each of the four `PairBuilder` subclasses. These instances are then used to create four instances of the `Extract` class. These four `Extract` objects can then build four `XYPair` objects from each source row.

The `dataclass.asdict()` function can be used to convert an `XYPair` object into a `dict[str, str]` object. This can be serialized by `json.dumps()` and written to an appropriate output file. This conversion operation seems like a good choice for a method in the abstract `PairBuilder` class. This can be used to write an `XYPair` object to an open file.

The top-level functions, `main()` and `get_options()`, can be placed in a separate module, named `acquisition`. This module will import the various class definitions from the `model` and `csv_extract` modules.

It's often helpful to review the SOLID design principles. In particular, we'll look closely at the **Dependency Inversion principle**.

Design principles

We can look at the SOLID design principles to be sure that the object-oriented design follows those principles.

- **Single Responsibility**: Each of the classes seems to have a single responsibility.

- **Open-Closed**: Each class seems open to extension by adding subclasses.

- **Liskov Substitution**: The `PairBuilder` class hierarchy follows this principle since each subclass is identical to the parent class.

- **Interface Segregation**: The interfaces for each class are minimized.

- **Dependency Inversion**: There's a subtle issue regarding dependencies among classes. We'll look at this in some detail.

One of the SOLID design principles suggests avoiding tight coupling between the `PairBuilder` subclasses and the `XYPair` class. The idea would be to provide a protocol (or interface) for the `XYPair` class. Using the protocol in type annotations would permit any type that implemented the protocol to be provided to the class. Using a protocol would break a direct dependency between the `PairBuilder` subclasses and the `XYPair` class.

This object-oriented design issue surfaces often, and generally leads to drawn-out, careful thinking about the relationships among classes and the SOLID design principles.

We have the following choices:

- Have a direct reference to the `XYPair` class inside the `PairBuilder` class. This would be `def from_row(row: list[str]) -> XYPair:`. This breaks the Dependency Inversion principle.

- Use `Any` as the type annotation. This would be
  `def from_row(row: list[str]) -> Any:`. This makes the type hints less informative.

- Attempt to create a protocol for the resulting type, and use this in the type hints.

- Introduce a type alias that (for now) only has one value. In future expansions of the model module, additional types might be introduced.

The fourth alternative gives us the flexibility we need for type annotation checking. The idea is to include a type alias like the following in the model module:

```
from dataclasses import dataclass
from typing import TypeAlias


@dataclass
class XYPair:
    # Definition goes here


RawData: TypeAlias = XYPair
```

As alternative classes are introduced, the definition of RawData can be expanded to include the alternatives. This might evolve to look like the following:

```
from dataclasses import dataclass
from typing import TypeAlias


@dataclass
class XYPair:
    # Definition goes here
    pass


@dataclass
class SomeOtherStructure:
    # Some other definition, here
    pass
```

```
RawData: TypeAlias = XYPair | SomeOtherStructure
```

This permits extension to the `PairBuilder` subclasses as the `model` module evolves. The `RawData` definition needs to be changed as new classes are introduced. Annotation-checking tools like **mypy** cannot spot the invalid use of any of the classes that comprise the alternative definitions of the `RawData` type alias.

Throughout the rest of the application, classes and functions can use `RawData` as an abstract class definition. This name represents a number of alternative definitions, any one of which might be used at run-time.

With this definition of `RawData`, the `PairBuilder` subclasses can use a definition of the following form:

```python
from model import RawData, XYPair
from abc import ABC, abstractmethod


class PairBuilder(ABC):
    target_class: type[RawData]

    @abstractmethod
    def from_row(self, row: list[str]) -> RawData:
        ...


class Series1Pair(PairBuilder):
    target_class = XYPair

    def from_row(self, row: list[str]) -> RawData:
        cls = self.target_class
        # the rest of the implementation...
        # return cls(arguments based on the value of row)
```

A similar analysis holds for the `main()` function. This can be directly tied to the `Extract` class and the various subclasses of the `PairBuilder` class. It's very important for these classes to be injected at run time, generally based on command-line arguments.

For now, it's easiest to provide the class names as default values. A function like the following might be used to get options and configuration parameters:

```
def get_options(argv: list[str]) -> argparse.Namespace:
    defaults = argparse.Namespace(
        extract_class=Extract,
        series_classes=[Series1Pair, Series2Pair, Series3Pair, Series4Pair],
    )
```

The `defaults` namespace is provided as an argument value to the `ArgumentParser.parse_args()` method. This set of defaults serves as a kind of dependency injection throughout the application. The `main` function can use these class names to build an instance of the given extract class, and then process the given source files.

A more advanced CLI could provide options and arguments to tailor the class names. For more complex applications, these class names would be read from a configuration file.

An alternative to the object-oriented design is a functional design. We'll look at that alternative in the next section.

Functional design

The general module structure shown in *Class design* applies to a functional design also. The `model` module with a single class definition is also a part of a functional design; this kind of module with a collection of dataclass definitions is often ideal.

As noted above in the *Design principles* section, the `model` module is best served by using a type variable, `RawData`, as a placeholder for any additional types that may be developed.

The `csv_extract` module will use a collection of independent functions to build `XYPair` objects. Each function will be similar in design.

Here are some example functions with type annotations:

```python
def series_1_pair(row: list[str]) -> RawData:
    ...


def series_2_pair(row: list[str]) -> RawData:
    ...


def series_3_pair(row: list[str]) -> RawData:
    ...


def series_4_pair(row: list[str]) -> RawData:
    ...
```

These functions can then be used by an `extract()` function to create the `XYPair` objects for each of the four series represented by a single row of the source file.

One possibility is to use the following kind of definition:

```python
SeriesBuilder: TypeVar = Callable[[list[str]], RawData]


def extract(row: list[str], builders: list[SeriesBuilder]) -> list[RawData]:
    ...
```

This `extract()` function can then apply all of the given builder functions (`series_1_pair()` to `series_4_pair()`) to the given row to create `XYPair` objects for each of the series.

This design will also require a function to apply `dataclass.asdict()` and `json.dumps()` to convert `XYPair` objects into strings that can be written to an NDJSON file.

Because the functions used are provided as argument values, there is little possibility of a dependency issue among the various functions that make up the application. The point throughout the design is to avoid binding specific functions in arbitrary places. The `main()` function should provide the row-building functions to the `extract` function. These functions can be provided via command-line arguments, a configuration file, or be default values if no overrides are given.

We've looked at the overall objective of the project, and two suggested architectural approaches. We can now turn to the concrete list of deliverables.

Deliverables

This project has the following deliverables:

- Documentation in the `docs` folder.

- Acceptance tests in the `tests/features` and `tests/steps` folders.

- Unit tests for model module classes in the `tests` folder.

- Mock objects for the `csv_extract` module tests will be part of the unit tests.

- Unit tests for the `csv_extract` module components in the `tests` folder.

- Application to acquire data from a CSV file in the `src` folder.

An easy way to start is by cloning the project zero directory to start this project. Be sure to update the `pyproject.toml` and `README.md` when cloning; the author has often been confused by out-of-date copies of old projects' metadata.

We'll look at a few of these deliverables in a little more detail. We'll start with some suggestions for creating the acceptance tests.

Acceptance tests

The acceptance tests need to describe the overall application's behavior from the user's point of view. The scenarios will follow the UX concept of a command-line application that acquires data and writes output files. This includes success as well as useful output in the event of failure.

The features will look something like the following:

```
Feature: Extract four data series from a file with
the peculiar Anscombe Quartet format.
```

```
Scenario: When requested, the application extracts all four series.
  Given the "Anscombe_quartet_data.csv" source file exists
  And the "quartet" directory exists
  When we run
    command "python src/acquire.py -o quartet Anscombe_quartet_data.csv"
  Then the "quartet/series_1.json" file exists
  And the "quartet/series_2.json" file exists
  And the "quartet/series_3.json" file exists
  And the "quartet/series_3.json" file exists
  And the "quartet/series_1.json" file starts with
    '{"x": "10.0", "y": "8.04"}'
```

This more complex feature will require several step definitions. These include the following:

- `@given('The "{name}" source file exists')`. This function should copy the example file from a source data directory to the temporary directory used to run the test.

- `@given('the "{name}" directory exists')`. This function can create the named directory under the directory used to run the test.

- `@then('the "{name}" file exists')`. This function can check for the presence of the named file in the output directory.

- `@then('the "quartet/series_1.json" file starts with ...')`. This function will examine the first line of the output file. In the event the test fails, it will be helpful to display the contents of the file to help debug the problem. A simple `assert` statement might not be ideal; a more elaborate `if` statement is needed to write debugging output and raise an `AssertionError` exception.

Because the application under test consumes and produces files, it is best to make use of the **behave** tool's `environment.py` module to define two functions to create (and destroy) a temporary directory used when running the test. The following two functions are used by **behave** to do this:

- `before\_scenario(context, scenario)`: This function can create a directory. The `tempfile` module has a `mkdtemp()` function that handles this. The directory needs to be placed into the context so it can be removed.

- `after_scenario(context, scenario)`: This function can remove the temporary directory.

The format for one of the `Then` clauses has a tiny internal inconsistency. The following uses a mixture of " and ' to make it clear where values are inserted into the text:

```
And the "quartet/series_1.json" file starts with'{"x": "10.0", "y": "8.04"}'
```

Some people may be bothered by the inconsistency. One choice is to use ' consistently. When there aren't too many feature files, this pervasive change is easy to make. Throughout the book, we'll be inconsistent, leaving the decision to make changes for consistency up to you.

Also, note the `When` clause command is rather long and complicated. The general advice when writing tests like this is to use a summary of the command and push the details into the step implementation function. We'll address this in a later chapter when the command becomes even longer and more complicated.

In addition to the scenario where the application works correctly, we also need to consider how the application behaves when there are problems. In the next section, we'll touch on the various ways things might go badly, and how the application should behave.

Additional acceptance scenarios

The suggested acceptance test covers only one scenario. This single scenario — where everything works — can be called the "happy path". It would be wise to include scenarios in which various kinds of errors occur, to be sure the application is reliable and robust in the face of problems. Here are some suggested error scenarios:

- Given the `Anscombe\_quartet\_data.csv` source file does not exist.
- Given the `quartet` directory does not exist.

- When we run the command `python src/acquire.py --unknown` option

- Given an `Anscombe\_quartet\_data.csv` source file exists, and the file is in the wrong format. There are numerous kinds of formatting problems.

    - The file is empty.

    - The file is not a proper CSV file, but is in some other format.

    - The file's contents are in valid CSV format, but the column names do not match the expected column names.

Each of the unhappy paths will require examining the log file to be sure it has the expected error messages. The **behave** tool can capture logging information. The `context` available in each step function has attributes that include captured logging output. Specifically, `context.log_capture` contains a `LogCapture` object that can be searched for an error message.

See `https://behave.readthedocs.io/en/stable/api.html#behave.runner.Context` for the content of the context.

These unhappy path scenarios will be similar to the following:

```
Scenario: When the file does not exist, the log has the expected
error message.
  Given the "Anscombe_quartet_data.csv" source file does not exist
  And the "quartet" directory exists
  When we run command "python src/acquire.py -o quartet
Anscombe_quartet_data.csv"
  Then the log contains "File not found: Anscombe_quartet_data.csv"
```

This will also require some new step definitions to handle the new `Given` and `Then` steps.

 When working with Gherkin, it's helpful to establish clear language and consistent terminology. This can permit a few step definitions to work for a large number of scenarios. It's a common experience to recognize similarities after writing several scenarios, and then choose to alter scenarios to simplify and normalize steps.

The **behave** tool will extract missing function definitions. The code snippets can be copied and pasted into a steps module.

Acceptance tests cover the application's overall behavior. We also need to test the individual components as separate units of code. In the next section, we'll look at unit tests and the mock objects required for those tests.

Unit tests

There are two suggested application architectures in *Architectural approach*. Class-based design includes two functions and a number of classes. Each of these classes and functions should be tested in isolation.

Functional design includes a number of functions. These need to be tested in isolation. Some developers find it easier to isolate function definitions for unit testing. This often happens because class definitions may have explicit dependencies that are hard to break.

We'll look at a number of the test modules in detail. We'll start with tests for the model module.

Unit testing the model

The model module only has one class, and that class doesn't really do very much. This makes it relatively easy to test. A test function something like the following should be adequate:

```python
from unittest.mock import sentinel
from dataclasses import asdict


def test_xypair():
    pair = XYPair(x=sentinel.X, y=sentinel.Y)
```

```
assert pair.x == sentinel.X
assert pair.y == sentinel.Y
assert asdict(pair) == {"x": sentinel.X, "y": sentinel.Y}
```

This test uses the `sentinel` object from the `unittest.mock` module. Each `sentinel` attribute — for example, `sentinel.X` — is a unique object. They're easy to provide as argument values and easy to spot in results.

In addition to testing the `model` module, we also need to test the `csv_extract` module, and the overall `acquire` application. In the next section, we'll look at the extract unit test cases.

Unit testing the PairBuilder class hierarchy

When following an object-oriented design, the suggested approach is to create a `PairBuilder` class hierarchy. Each subclass will perform slightly different operations to build an instance of the `XYPair` class.

Ideally, the implementation of the `PairBuilder` subclasses is not tightly coupled to the `XYPair` class. There is some advice in the *Design principles* section on how to support dependency injection via type annotations. Specifically, the `model` module is best served by using a type variable, `RawData`, as a placeholder for any additional types that may be developed.

When testing, we want to replace this class with a mock class to assure that the interface for the family of `RawData` classes — currently only a single class, `XYPair` — is honored.

A `Mock` object (built with the `unittest.mock` module) works out well as a replacement class. It can be used for the `XYPair` class in the subclasses of the `PairBuilder` class.

The tests will look like the following example:

```
from unittest.mock import Mock, sentinel, call


def test_series1pair():
    mock_raw_class = Mock()
    p1 = Series1Pair()
```

```
p1.target_class = mock_raw_class
xypair = p1.from_row([sentinel.X, sentinel.Y])
assert mock_raw_class.mock_calls == [
    call(sentinel.X, sentinel.Y)
]
```

The idea is to use a Mock object to replace the specific class defined in the Series1Pair class. After the from_row() method is evaluated, the test case confirms that the mock class was called exactly once with the expected two sentinel objects. A further check would confirm that the value of xypair was also a mock object.

This use of Mock objects guarantees that no additional, unexpected processing was done on the objects. The interface for creating a new XYPair was performed correctly by the Series1Pair class.

Similar tests are required for the other pair-building classes.

In addition to testing the model and csv_extract modules, we also need to test the overall acquire application. In the next section, we'll look at the acquire application unit test cases.

Unit testing the remaining components

The test cases for the overall Extract class will also need to use Mock objects to replace components like a csv.reader and instances of the PairBuilder subclasses.

As noted above in the *Functional design* section, the main() function needs to avoid having explicitly named classes or functions. The names need to be provided via command-line arguments, a configuration file, or as default values.

The unit tests should exercise the main() function with Mock objects to be sure that it is defined with flexibility and extensions in mind.

Summary

This chapter introduced the first project, the Data Acquisition Base Application. This application extracts data from a CSV file with a complex structure, creating four separate series of data points from a single file.

To make the application complete, we included a command-line interface and logging. This will make sure the application behaves well in a controlled production environment.

An important part of the process is designing an application that can be extended to handle data from a variety of sources and in a variety of formats. The base application contains modules with very small implementations that serve as a foundation for making subsequent extensions.

Perhaps the most difficult part of this project is creating a suite of acceptance tests to describe the proper behavior. It's common for developers to compare the volume of test code with the application code and claim testing is taking up "too much" of their time.

Pragmatically, a program without automated tests cannot be trusted. The tests are every bit as important as the code they're exercising.

The unit tests are — superficially — simpler. The use of mock objects makes sure each class is tested in isolation.

This base application acts as a foundation for the next few chapters. The next chapter will add RESTful API requests. After that, we'll have database access to this foundation.

Extras

Here are some ideas for you to add to this project.

Logging enhancements

We skimmed over logging, suggesting only that it's important and that the initialization for logging should be kept separate from the processing within the `main()` function.

The `logging` module has a great deal of sophistication, however, and it can help to explore this. We'll start with logging "levels".

Many of our logging messages will be created with the INFO level of logging. For example:

```
logger.info("%d rows processed", input_count)
```

This application has a number of possible error situations that are best reflected with **error**-level logging.

Additionally, there is a tree of named loggers. The root logger, named " ", has settings that apply to all the lower-level loggers. This tree tends to parallel the way object inheritance is often used to create classes and subclasses. This can make it advantageous to create loggers for each class. This permits setting the logging level to **debug** for one of many classes, allowing for more focused messages.

This is often handled through a logging configuration file. This file provides the configuration for logging, and avoids the potential complications of setting logging features through command-line options.

There are three extras to add to this project:

- Create loggers for each individual class.

- Add debug-level information. For example, the from_row() function is a place where debugging might be helpful for understanding why an output file is incorrect.

- Get the logging configuration from an initialization file. Consider using a file in **TOML** format as an alternative to the **INI** format, which is a first-class part of the logging module.

Configuration extensions

We've described a little of the CLI for this application. This chapter has provided a few examples of the expected behavior. In addition to command-line parameters, it can help to have a configuration file that provides the slowly changing details of how the application works.

In the discussion in the *Design principles* section, we looked closely at dependency inversion. The intent is to avoid an explicit dependency among classes. We want to "invert" the relationship, making it indirect. The idea is to inject the class name at run time, via parameters.

Initially, we can do something like the following:

```
EXTRACT_CLASS: type[Extract] = Extract
BUILDER_CLASSES: list[type[PairBuilder]] = [
    Series1Pair, Series2Pair, Series3Pair, Series4Pair]


def main(argv: list[str]) -> None:
    builders = [cls() for vls in BUILDER_CLASSES]
    extractor = EXTRACT_CLASS(builders)
    # etc.
```

This provides a base level of parameterization. Some global variables are used to "inject" the run-time classes. These initializations can be moved to the `argparse.Namespace` initialization value for the `ArgumentParser.parse_args()` method.

The initial values for this `argparse.Namespace` object can be literal values, essentially the same as shown in the global variable parameterization shown in the previous example.

It is more flexible to have the initial values come from a parameter file that's separate from the application code. This permits changing the configuration without touching the application and introducing bugs through inadvertent typing mistakes.

There are two popular alternatives for a configuration file that can be used to fine-tune the application. These are:

- A separate Python module that's imported by the application. A module name like `config.py` is popular for this.

- A non-Python text file that's read by the application. The TOML file format, parsed by the `tomllib` module, is ideal.

Starting with Python 3.11, the `tomllib` module is directly available as part of the standard library. Older versions of Python should be upgraded to 3.11 or later.

When working with a TOML file, the class name will be a string. The simple and reliable way to translate the class name from string to class object is to use the `eval()` function. An alternative is to provide a small dictionary with class name strings and class objects. Class names can be resolved through this mapping.

> Some developers worry that the `eval()` function allows a class of Evil Super Geniuses to tweak the configuration file in a way that will crash the application.
>
> What these developers fail to notice is that the entire Python application is plain text. The Evil Super Genius can more easily edit the application and doesn't need to do complicated, nefarious things to the parameter file.
>
> Further, ordinary OS-level ownership and permissions can restrict access to the parameter file to a few trustworthy individuals.

Don't forget to include unit test cases for parsing the parameter file. Also, an acceptance test case with an invalid parameter file will be an important part of this project.

Data subsets

To work with large files it will be necessary to extract a subset of the data. This involves adding features like the following:

- Create a subclass of the `Extract` class that has an upper limit on the number of rows created. This involves a number of unit tests.

- Update the CLI options to include an optional upper limit. This, too, will involve some additional unit test cases.

- Update the acceptance test cases to show operation with the upper limit.

Note that switching from the `Extract` class to the `SubsetExtract` class is something that should be based on an optional command-line parameter. If the `--limit` option is not given, then the `Extract` class is used. If the `--limit` option is given (and is a valid integer), then the `SubsetExtract` class is used. This will lead to an interesting set of unit test cases to make sure the command-line parsing works properly.

Another example data source

Perhaps the most important extra for this application is to locate another data source that's of interest to you.

See the **CO$_2$ PPM — Trends in Atmospheric Carbon Dioxide** data set, available at `https://datahub.io/core/co2-ppm`, for some data that's somewhat larger. This has a number of odd special-case values that we'll explore in *Chapter 6, Project 2.1: Data Inspection Notebook.*

This project will require you to manually download and unzip the file. In later chapters, we'll look at automating these two steps. See *Chapter 4, Data Acquisition Features: Web APIs and Scraping* specifically, for projects that will expand on this base project to properly acquire the raw data from a CSV file.

What's important is locating a source of data that's in CSV format and small enough that it can be processed in a few seconds. For large files, it will be necessary to extract a subset of the data. See *Data subsets* for advice on handling large sets of data.

4

Data Acquisition Features: Web APIs and Scraping

Data analysis often works with data from numerous sources, including databases, web services, and files prepared by other applications. In this chapter, you will be guided through two projects to add additional data sources to the baseline application from the previous chapter. These new sources include a web service query, and scraping data from a web page.

This chapter's projects cover the following essential skills:

- Using the **requests** package for Web API integration. We'll look at the Kaggle API, which requires signing up to create an API token.

- Using the **Beautiful Soup** package to parse an HTML web page.

- Adding features to an existing application and extending the test suite to cover these new alternative data sources.

It's important to recognize this application has a narrow focus on data acquisition. In later chapters, we'll validate the data and convert it to a more useful form. This reflects a separation of the following distinct concerns:

- Downloading and extracting data from the source are the foci of this chapter and the next.

- Inspection begins in *Chapter 6, Project 2.1: Data Inspection Notebook*.

- Validating and cleaning the data starts in *Chapter 9, Project 3.1: Data Cleaning Base Application*.

Each stage in the processing pipeline is allocated to separate projects. For more background, see *Chapter 2, Overview of the Projects*.

We'll start by looking at getting data using an API and a RESTful web service. This will focus on the Kaggle site, which means you will need to sign up with Kaggle to get your own, unique API key. The second project will scrape HTML content from a website that doesn't offer a useful API.

Project 1.2: Acquire data from a web service

It's common to need data provided by a Web API. One common design approach for web services is called RESTful; it's based on a number of concepts related to using the HTTP protocol to transfer a representation of an object's state.

For more information on RESTful services, see *Building RESTful Python Web Services* (`https://www.packtpub.com/product/building-restful-python-web-services/9781 786462251`).

A RESTful service generally involves using the HTTP protocol to respond to requests from client applications. The spectrum of request types includes a number of verbs like get, post, put, patch, and delete. In many cases, the service responds with a JSON document. It's also possible to receive a file that's a stream of NDJSON documents, or even a file that's a ZIP archive of data.

We'll start with a description of the application, and then move on to talk about the architectural approach. This will be followed with a detailed list of deliverables.

Description

Analysts and decision-makers need to acquire data for further analysis. In this case, the data is available from a RESTful web service. One of the most fun small data sets to work with is Anscombe's Quartet — `https://www.kaggle.com/datasets/carlmcbrideellis/data-anscombes-quartet`

Parts of this application are an extension to the project in *Chapter 9, Project 3.1: Data Cleaning Base Application*. The essential behavior of this application will be similar to the previous project. This project will use a CLI application to grab data from a source.

The **User Experience (UX)** will also be a command-line application with options to fine-tune the data being gathered. Our expected command line should like something like the following:

```
% python src/acquire.py -o quartet -k ~/Downloads/kaggle.json \
    --zip carlmcbrideellis/data-anscombes-quartet
```

The `-o quartet` argument specifies a directory into which four results are written. These will have names like `quartet/series_1.json`.

The `-k kaggle.json` argument is the name of a file with the username and Kaggle API token. This file is kept separate from the application software. In the example, the file was in the author's `Downloads` folder.

The `--zip` argument provides the "reference" — the owner and data set name — to open and extract. This information is found by examining the details of the Kaggle interface.

An additional feature is to get a filtered list of Kaggle data sets. This should be a separate `--search` operation that can be bundled into a single application program.

```
% python src/acquire.py --search -k ~/Downloads/kaggle.json
```

This will apply some search criteria to emit a list of data sets that match the requirements. The lists tend to be quite large, so this needs to be used with care.

The credentials in the file are used to make the Kaggle API request. In the next sections, we'll look at the Kaggle API in general. After that, we'll look at the specific requests required to locate the reference to the target data set.

The Kaggle API

See `https://www.kaggle.com/docs/api` for information on the Kaggle API. This document describes some command-line code (in Python) that uses the API.

The technical details of the RESTful API requests are at `https://github.com/Kaggle/ka ggle-api/blob/master/KaggleSwagger.yaml`. This document describes the requests and responses from the Kaggle API server.

To make use of the RESTful API or the command-line applications, you should register with Kaggle. First, sign up with `Kaggle.com`. Next, navigate to the public profile page. On this page, there's an API section. This section has the buttons you will use to generate a unique API token for your registered username.

The third step is to click the **Create New Token** button to create the token file. This will download a small JSON file with your registered username and unique key. These credentials are required by the Kaggle REST API.

The ownership of this file can be changed to read-only by the owner. In Linux and macOS, this is done with the following command:

```
% chmod 400 ~/Downloads/kaggle.json
```

Do not move the Kaggle credentials file named `kaggle.json` into a directory where your code is also located. It's tempting, but it's a terrible security mistake because the file could get saved to a code repository and become visible to anyone browsing your code. In some enterprises, posting keys in code repositories — even internal repositories — is a security lapse and a good reason for an employee to be terminated.

Because Git keeps a very complete history, it's challenging to remove a commit that contains keys.

Keep the credentials file separate from your code.

It's also a good idea to add `kaggle.json` to a `.gitignore` file to make extra sure that it won't be uploaded as part of a commit and push.

About the source data

This project will explore two separate kinds of source data. Both sources have the same base path of `https://www.kaggle.com/api/v1/`. Trying to query this base path won't provide a useful response; it's only the starting point for the paths that are built to locate specific resources.

- JSON documents with summaries of data sets or metadata about data sets. These come from appending `datasets/list` to the base path.

- A ZIP archive that contains the data we'll use as an example. This comes from appending `datasets/download/{ownerSlug}/{datasetSlug}` to the base path. The `ownerSlug` value is "carlmcbrideellis". The `datasetSlug` value is "data-anscombes-quartet". A given data set has a `ref` value as a reference string with the required "ownerSlug/datasetSlug" format.

The JSON documents require a function to extract a few relevant fields like `title`, `ref`, `url`, and `totalBytes`. This subset of the available metadata can make it easier to locate useful, interesting data sets. There are numerous other properties available for search, like `usabilityRating`; these attributes can distinguish good data sets from experiments or classroom work.

The suggested data set — the Anscombe Quartet — is available as a ZIP-compressed archive with a single item inside it. This means the application must handle ZIP archives and expand a file contained within the archive. Python offers the `zipfile` package to handle locating the CSV file within the archive. Once this file is found, the existing programming from the previous chapter (*Chapter 3, Project 1.1: Data Acquisition Base Application*) can be used.

There are thousands of Kaggle data sets. We'll suggest some alternatives to the Anscombe Quartet in the *Extras*.

This section has looked at the input, processing, and output of this application. In the next section, we'll look at the overall architecture of the software.

Approach

We'll take some guidance from the C4 model (`https://c4model.com`) when looking at our approach:

- **Context**: For this project, a context diagram would show a user extracting data from a source. You may find it helpful to draw this diagram.

- **Containers**: One container is the user's personal computer. The other container is the Kaggle website, which provides the data.

- **Components**: We'll address the components below.

- **Code**: We'll touch on this to provide some suggested directions.

It's important to consider this application as an extension to the project in *Chapter 3, Project 1.1: Data Acquisition Base Application*. The base level of architectural design is provided in that chapter.

In this project, we'll be adding a new `kaggle_client` module to download the data. The overall application in the `acquire` module will change to make use of this new module. The other modules should remain unchanged.

The legacy component diagram is shown in *Figure 4.1*.

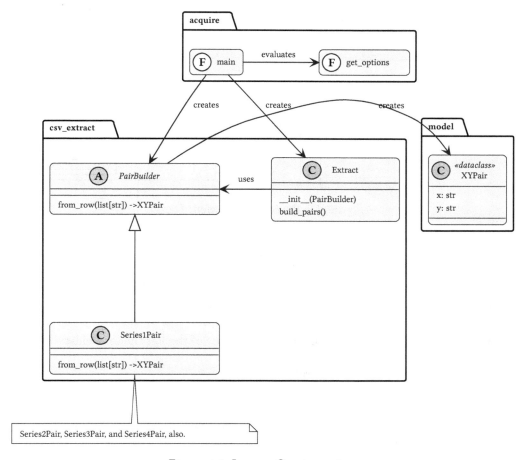

Figure 4.1: Legacy Components

A new architecture can handle both the examination of the JSON data set listing, as well as the download of a single ZIP file. This is shown in *Figure 4.2*.

The new module here is the `kaggle_client` module. This has a class, `RestAccess`, that provides methods to access Kaggle data. It can reach into the Kaggle data set collection and retrieve a desired ZIP file. Additional methods can be added to examine the list of data

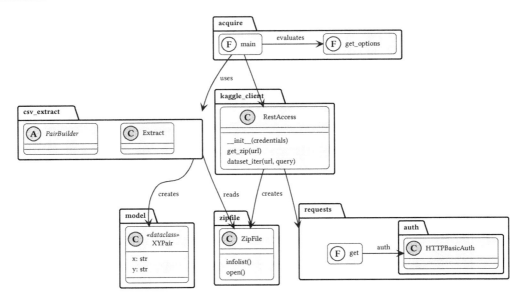

Figure 4.2: Revised Component Design

sets or get data set metadata.

The RestAccess class is initialized with the contents of the kaggle.json file. As part of initialization, it can create the required authentication object for use in all subsequent calls.

In the following sections, we'll look at these features of the RestAccess class:

- Making API requests in general.

- Getting the ZIP archive.

- Getting the list of data sets.

- Handling the rate-limiting response.

We'll start with the most important feature, making API requests in a general way.

Making API requests

The component diagram shows the requests package as the preferred way to access RESTful APIs. This package should be added to the project's pyproject.toml and installed as part of the project's virtual environment.

It's also sensible to make RESTful API requests with the `urllib` package. This is part of the standard library. It works nicely and requires no additional installation. The code can become rather complicated-looking, however, so it's not as highly recommended as the `requests` package.

The essential benefit of using `requests` is creating an authentication object and providing it in each request. We often use code like the following example:

```python
import json
from pathlib import Path
import requests.auth


keypath = Path.home() / "Downloads" / "kaggle.json"
with keypath.open() as keyfile:
    credentials = json.load(keyfile)
auth = requests.auth.HTTPBasicAuth(
    credentials['username'], credentials['key']
)
```

This can be part of the __init__() method of the `RestAccess` class.

The auth object created here can be used to make all subsequent requests. This will provide the necessary username and API token to validate the user. This means other methods can, for example, use `requests.get()` with a keyword parameter value of `auth=self.auth`. This will correctly build the needed `Authorization` headers in each request.

Once the class is initialized properly, we can look at the method for downloading a ZIP archive

Downloading a ZIP archive

The `RestAccess` class needs a `get_zip()` method to download the ZIP file. The parameter is the URL for downloading the requested data set.

The best approach to building this URL for this data set is to combine three strings:

- The base address for the APIs, `https://www.kaggle.com/api/v1`.

- The path for downloads, `/datasets/download/`.

- The reference is a string with the form: `{ownerSlug}/{datasetSlug}`.

This is an ideal place for a Python f-string to replace the reference in the URL pattern.

The output from the `get_zip()` method should be a `Path` object. In some cases, the ZIP archives are gigantic and can't be processed entirely in memory. In these extreme cases, a more complicated, chunked download is required. For these smaller files used by this project, the download can be handled entirely in memory. Once the ZIP file has been written, the client of this `RestAccess` class can then open it and extract the useful member.

A separate client function or class will process the content of the ZIP archive file. The following is **not** part of the `RestAccess` class but is part of some client class or function that uses the `RestAccess` class.

Processing an element of an archive can be done with two nested `with` contexts. They would work like this:

- An outer `with` statement uses the `zipfile` module to open the archive, creating a `ZipFile` instance.

- An inner `with` statement can open the specific member with the Anscombe quartet CSV file. Inside this context, the application can create a `csv.DictReader` and use the existing `Extract` class to read and process the data.

What's important here is we don't need to unpack the ZIP archive and litter our storage with unzipped files. An application can open and process the elements using the `ZipFile.open()` method.

In addition to downloading the ZIP archive, we may also want to survey the available data sets. For this, a special iterator method is helpful. We'll look at that next.

Getting the data set list

The catalog of data sets is found by using the following path:

```
https://www.kaggle.com/api/v1/datasets/list
```

The `RestAccess` class can have a `dataset_iter()` method to iterate through the collection of data sets. This is helpful for locating other data sets. It's not required for finding the Anscombe's Quartet, since the `ownerSlug` and `datasetSlug` reference information is already known.

This method can make a `GET` request via the `requests.get()` function to this URL. The response will be on the first page of available Kaggle data sets. The results are provided in pages, and each request needs to provide a page number as a parameter to get subsequent pages.

Each page of results will be a JSON document that contains a sequence of dictionary objects. It has the following kind of structure:

```
[
    {"id": some_number, "ref": "username/dataset", "title": ...},
    {"id": another_number, "ref": "username/dataset", "title": ...},
    etc.
]
```

This kind of two-tiered structure — with pages and items within each page — is the ideal place to use a generator function to iterate through the pages. Within an outer cycle, an inner iteration can yield the individual data set rows from each page.

This nested iteration can look something like the following code snippet:

```
def dataset_iter(url: str, query: dict[str, str]) ->
  Iterator[dict[str, str]]:
    page = 1
    while True:
        response = requests.get(url, params=quert | {"page": str(page)})
        if response.status_code == 200:
            details = response.json()
```

```
        if details:

            yield from iter(details)

            page += 1

        else:

            break

    elif response.status_code == 429:

        # Too Many Requests

        # Pause and try again processing goes here...

        pass

    else:

        # Unexpected response

        # Error processing goes here...

        break
```

This shows the nested processing of the `while` statement ends when a response contains a page of results with zero entries in it. The processing to handle too many requests is omitted. Similarly, the logging of unexpected responses is also omitted.

A client function would use the `RestAccess` class to scan data sets and would look like the following example:

```
keypath = Path.home()/"Downloads"/"kaggle.json"
with keypath.open() as keyfile:
    credentials = json.load(keyfile)


reader = Access(credentials)
for row in reader.dataset_iter(list_url):
    print(row['title'], row['ref'], row['url'], row['totalBytes'])
```

This will process all of the data set descriptions returned by the `RestReader` object, `reader`. The `dataset_iter()` method needs to accept a `query` parameter that can limit the scope of the search. We encourage you to read the OpenAPI specification to see what options are possible for the `query` parameter. These values will become part of the query string in the HTTP `GET` request.

Here's the formal definition of the interface:

`https://github.com/Kaggle/kaggle-api/blob/master/KaggleSwagger.yaml`

Some of the query parameters include the following:

- The `filetype` query is helpful in locating data in JSON or CSV formats.

- The `maxSize` query can constrain the data sets to a reasonable size. For initial exploration, 1MB is a good upper limit.

Initial spike solutions — without regard to the rate limiting — will turn up at least 80 pages of possible data sets. Handling the rate-limiting response will produce more extensive results, at the cost of some time spent waiting. In the next section, we'll expand this method to handle the error response.

Rate limiting

As with many APIs, the Kaggle API imposes rate-limiting to avoid a **Denial-of-Service (DoS)** attack. For more information see `https://cheatsheetseries.owasp.org/cheatsheets/Denial_of_Service_Cheat_Sheet.html`.

Each user has a limited number of requests per second. While the limit is generous for most purposes, it will tend to prevent a simple scan of **all** data sets.

A status code of 429 in a Kaggle response tells the client application that too many requests were made. This "too many requests" error response will have a header with the key `Retry-After`. This header's value is the timeout interval (in seconds) before the next request can be made.

A reliable application will have a structure that handles the 429 vs. 200 responses gracefully. The example in the previous section has a simple `if` statement to check the condition `if response.status_code == 200`. This needs to be expanded to handle these three alternatives:

- A status code of 200 is a good response. If the page has any details, these can be processed; the value of page can be incremented. Otherwise, there's no more data making it appropriate to break from the containing while statement.

- A status code of 429 means too many requests were made. Get the value of the Retry-After and sleep for this period of time.

- Any other status code indicates a problem and should be logged or raised as an exception.

One possible algorithm for return rows and handling rate limiting delays is shown in *Figure 4.3.*

Handling rate limiting will make the application much easier to use. It will also produce more complete results. Using an effective search filter to reduce the number of rows to a sensible level will save a lot of waiting for the retry-after delays.

The main() function

The current application design has these distinct features:

1. Extract data from a local CSV file.

2. Download a ZIP archive and extract data from a CSV member of the archive.

3. (Optionally) Survey the data set list to find other interesting data sets to process.

This suggests that our main() function should be a container for three distinct functions that implement each separate feature. The main() function can parse the command-line arguments, and then make a number of decisions:

- **Local Extract**: If the -o option is present (without the -k option), then this is a local file extract. This was the solution from an earlier chapter.

- **Download and Extract**: If the -k and -o options are present, then this will be a download and extract. It will use the RestAccess object to get the ZIP archive. Once the archive is opened, the member processing is the solution from an earlier chapter.

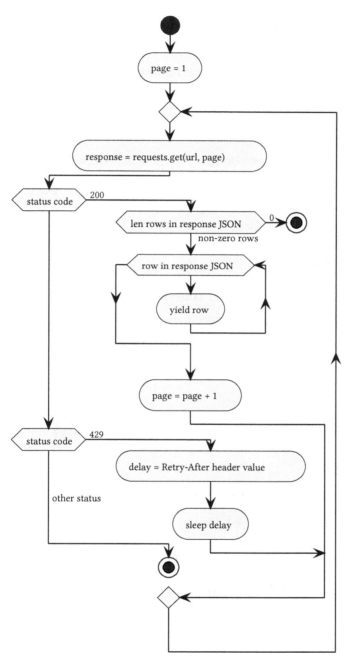

Figure 4.3: Kaggle rate-limited paging

- **Survey**: If the `-k` and `-s` (or `--search`) options are present, then this is a search for interesting data sets. You are encouraged to work out the argument design to provide the needed query parameters to the application.

- **Otherwise**: If none of the above patterns match the options, this is incoherent, and an exception should be raised.

Each of these features requires a distinct function. A common alternative design is to use the **Command** pattern and create a class hierarchy with each of the features as a distinct subclass of some parent class.

One central idea is to keep the `main()` function small, and dispatch the detailed work to other functions or objects.

The other central idea is **Don't Repeat Yourself** (**DRY**). This principle makes it imperative to **never** copy and paste code between the "Download-and-Extract" feature and the "Local-Extract" feature. The "Download-and-Extract" processing must reuse the "Local-Extract" processing either through subclass inheritance or calling one function from another.

Now that we have a technical approach, it's time to look at the deliverables for this project.

Deliverables

This project has the following deliverables:

- Documentation in the `docs` folder.

- Acceptance tests in the `tests/features` and `tests/steps` folders.

- A miniature RESTful web service that provides test responses will be part of the acceptance test.

- Unit tests for the application modules in the `tests` folder.

- Mock objects for the `requests` module will be part of the unit tests.

- Application to download and acquire data from a RESTful web service.

Be sure to include additional packages like `requests` and `beautifulsoup4` in the `pyproject.toml` file. The **pip-compile** command can be used to create a `requirements.txt` usable by the **tox** tool for testing.

We'll look at a few of these deliverables in a little more detail.

Unit tests for the RestAccess class

For unit testing, we don't want to involve the `requests` module. Instead, we need to make a mock interface for the `requests` module to confirm the application `RestAccess` module uses the `requests` classes properly.

There are two strategies for plugging in mock objects:

- Implement a dependency injection technique, where the target class is named at run time.

- Use *Monkey Patching* to inject a mock class at test time.

When working with external modules — modules where we don't control the design — monkey patching is often easier than trying to work out a dependency injection technique. When we're building the classes in a module, we often have a need to extend the definitions via subclasses. One of the reasons for creating unique, customized software is to implement change in the unique features of an application rapidly. The non-unique features (RESTful API requests, in this case) change very slowly and don't benefit from flexibility.

We want to create two mock classes, one to replace the `requests.auth.HTTPBasicAuth` class, and one to replace the `requests.get()` function. The mock for the `HTTPBasicAuth` class doesn't do anything; we want to examine the mock object to be sure it was called once with the proper parameters. The mock for the `requests.get()` function needs to create mock `Response` objects for various test scenarios.

We'll need to use the `monkeypatch` fixture of the `pytest` module to replace the real objects with the mock objects for unit testing.

The idea is to create unit tests that have a structure similar to the following example:

```python
from unittest.mock import Mock, sentinel, call


def test_rest_access(monkeypatch):
    mock_auth_class = Mock(
        name="Mocked HTTPBasicAuth class",
        return_value=sentinel.AUTH
    )
    monkeypatch.setattr('requests.auth.HTTPBasicAuth', mock_auth_class)
    mock_kaggle_json = {"username": sentinel.USERNAME, "key": sentinel.KEY}
    access = RestAccess(mock_kaggle_json)
    assert access.credentials == sentinel.AUTH
    assert mock_auth_class.mock_calls == [
        call(sentinel.USERNAME, sentinel.KEY)
    ]
```

This test case creates a mock for the HTTPBasicAuth class. When the class is called to create an instance, it returns a sentinel object that can be verified by a test case.

The monkeypatch fixture replaces the requests.auth.HTTPBasicAuth class with the mock object. After this patch is applied, when the RestAccess class initialization attempts to create an instance of the HTTPBasicAuth class, it will invoke the mock, and will get a sentinel object instead.

The case confirms the sentinel object is used by the RestAccess instance. The test case also confirms the mock class was called exactly once with the expected values taken from the mocked value loaded from the kaggle.json file.

This test case relies on looking inside the RestAccess instance. This isn't the best strategy for writing unit tests. A better approach is to provide a mock object for requests.get(). The test case should confirm the requests.get() is called with a keyword parameter, auth, with an argument value of the sentinel.AUTH object. The idea of this test strategy is to examine the external interfaces of the RestAccess class instead of looking at internal state changes.

Acceptance tests

The acceptance tests need to rely on a *fixture* that mocks the Kaggle web service. The mock will be a process on your local computer, making it easy to stop and start the mock service to test the application. Using an address of 127.0.0.1:8080 instead of www.kaggle.com will direct RESTful API requests back to your own computer. The name localhost:8080 can be used instead of the numeric address 127.0.0.1:8080. (This address is called the *Loopback Address* because the requests loop back to the same host that created them, allowing testing to proceed without any external network traffic.)

Note that the URL scheme will change to http: from https:, also. We don't want to implement the full Socket Security Layer (SSL) for acceptance testing. For our purposes, we can trust those components work.

This change to the URLs suggests the application should be designed in such a way to have the https://www.kaggle.com portion of each URL provided by a configuration parameter. Then acceptance tests can use http://127.0.0.1:8080 without having to make any changes to the code.

The mock service must offer a few features of the Kaggle service. The local service needs to respond to dataset/download requests properly, providing a reply with the expected status code and the content with bytes that are a proper ZIP archive.

This mock service will run as a separate application. It will be started (and stopped) by **behave** for a scenario that needs the fixture.

We'll start by looking at the way this service is described in a feature file. This will lead us to look at how to build the mock service. After that, we can look at how this is implemented using **behave** step definitions.

The feature file

The downloading feature is clearly separate from the data acquisition feature. This suggests a new .feature file to provide scenarios to describe this feature.

Within this new feature file, we can have scenarios that specifically name the required fixture. A scenario might look like the following example:

```
@fixture.kaggle_server
Scenario: Request for carlmcbrideellis/data-anscombes-quartet
    extracts file from ZIP archive.
    A typical download command might be
    "python src/acquire.py -k kaggle.json -o quartet \
      --zip carlmcbrideellis/data-anscombes-quartet"


  Given proper keys are in "kaggle.json"
  When we run the kaggle download command
  Then log has INFO line with "header: ['mock', 'data']"
  And log has INFO line with "count: 1"
```

The `@fixture.` tag follows the common tagging convention for associating specific fixtures with scenarios. There are many other purposes for tagging scenarios in addition to specifying the fixture to use.

In previous projects, a command to run the application was provided in the `When` step. For this scenario (and many others), the command text became too long to be usefully presented in the Gherkin text. This means the actual command needs to be provided by the function that implements this step.

This scenario's `Then` steps look at the log created by the application to confirm the contents of the file.

A test scenario is part of the overall application's requirements and design.

In the description provided in *Description* there isn't any mention of a log. This kind of gap is common. The test scenario provided additional definitions of the feature omitted from the plain test description.

Some people like to update the documentation to be complete and fully consistent. We encourage flexibility when working on enterprise applications where there are numerous stakeholders. It can be difficult to get everyone's input into the initial presentation or document. Sometimes, requirements appear later in the process when more concrete issues, like expected operating scenarios, are discussed in detail.

The tag information will be used by the **behave** tool. We'll look at how to write a before_tag() function to start (and stop) the special mock server for any scenario that needs it.

Before we look at the **behave** integration via a step definition, we'll look at two approaches to testing the client application. The core concept is to create a mock-up of the few elements of the Kaggle API used by the data acquisition application. This mock-up must return responses used by test scenarios. There are two approaches:

- Create a web service application. For acceptance tests, this service must be started and stopped. The acquire application can be configured with a http://localhost:8080 URL to connect to the test server instead of the Kaggle server. (There are a few common variations on the "localhost" address including 127.0.0.1 and 0.0.0.0.)

- The other approach is to provide a way to replace the requests module with a mocked version of the module. This mocked module returns responses appropriate to the test scenario. This can be done by manipulating the sys.path variable to include the directory containing the mocked version of requests in front of the site-packages directory, which has the real version. It can also be done by providing some application configuration settings that can be replaced with the mocked package.

One approach to creating a complete service that will implement the fixture.

Injecting a mock for the requests package

Replacing the `requests` package requires using dependency injection techniques in the acquire application. A static dependency arises from code like the following:

```
import requests
import requests.auth
```

Later in the module, there may be code like `requests.get(...)` or `requests.auth.HTTPBasicAuth(...)`. The binding to the `requests` module is fixed by both the `import` statement and the references to `requests` and `requests.auth`.

The `importlib` module permits more dynamic binding of modules names, allowing some run-time flexibility. The following, for example, can be used to tailor imports.

```
if __name__ == "__main__":
    # Read from a configuration file
    requests_name = "requests"
    requests = importlib.import_module(requests_name)
    main(sys.argv[1:])
```

The global variable, `requests`, has the imported module assigned to it. This module variable **must** be global; it's an easy requirement to overlook when trying to configure the application for acceptance testing.

Note that the import of the `requests` module (or the mock version) is separated from the remaining `import` statements. This can be the source of some confusion for folks reading this code later, and suitable comments are important for clarifying the way this dependency injection works.

When we looked at unit testing in *Unit tests for the RestAccess class*, we used the **pytest** fixture named `monkeypatch` to properly isolate modules for testing.

Monkey patching isn't a great technique for acceptance testing because the code being tested is not **exactly** the code that will be used. While monkey patching and dependency injection are popular, there are always questions about testing patched software instead of the actual software. In some industries — particularly those where human lives might be at risk from computer-controlled machinery — the presence of a patch for testing may not be allowed. In the next section, we'll look at building a mock service to create and test the acquire application without any patching or changes.

Creating a mock service

A mock service can be built with any web services framework. There are two that are part of the standard library: the http.server package and the wsgiref package. Either of these can respond to HTTP requests, and can be used to create local services that can mock the Kaggle web service to permit testing our client.

Additionally, any of the well-known web service frameworks can be used to create the mock service. Using a tool like **flask** or **bottle** can make it slightly easier to build a suitable mock service.

To keep the server as simple as possible, we'll use the **bottle** framework. This means adding bottle==0.12.23 to the pyproject.toml file in the [project.optional-dependencies] section. This tool is only needed by developers.

The **Bottle** implementation of a RESTful API might look like this:

```python
import csv
import io
import json
import zipfile
from bottle import route, run, request, HTTPResponse


@route('/api/v1/datasets/list')
def datasets_list(name):
    # Provide mock JSON documents and Rate throttling


@route('/api/v1/datasets/download/<ownerSlug>/<datasetSlug>')
def datasets_download(ownerSlug, datasetSlug):
```

```
    # Provide mock ZIP archive


if __name__ == "__main__":
    run(host='127.0.0.1', port=8080)
```

While the Kaggle service has numerous paths and methods, this data acquisition project application doesn't use all of them. The mock server only needs to provide routes for the paths the application will actually use.

The `datasets_list` function might include the following example response:

```
@route('/api/v1/datasets/list')
def datasets_list(name):
    page = request.query.page or '1'
    if page == '1':
        mock_body = [
            # Provide attributes as needed by the application under test
            {'title': 'example1'},
            {'title': 'example2'},
        ]
        response = HTTPResponse(
            body=json.dumps(mock_body),
            status=200,
            headers={'Content-Type': 'application/json'}
        )
    else:
        # For error-recovery scenarios, this response may change.
        response = HTTPResponse(
            body=json.dumps([]),
            status=200,
            headers={'Content-Type': 'application/json'}
        )
```

```
    return response
```

The HTTPResponse object contains the essential features of the responses as seen by the acquisition application's download requests. Each response has content, a status code, and a header that is used to confirm the type of response.

For more comprehensive testing, it makes sense to add another kind of response with the status code of 429 and a header dictionary with {'Retry-After': '30'}. For this case, the two values of response will be more dramatically distinct.

The download needs to provide a mocked ZIP archive. This can be done as shown in the following example:

```python
@route('/api/v1/datasets/download/<ownerSlug>/<datasetSlug>')
def datasets_download(ownerSlug, datasetSlug):
    if ownerSlug == "carlmcbrideellis" and datasetSlug ==
      "data-anscombes-quartet":
        zip_content = io.BytesIO()
        with zipfile.ZipFile(zip_content, 'w') as archive:
            target_path = zipfile.Path(archive, 'Anscombe_quartet_data.csv')
            with target_path.open('w') as member_file:
                writer = csv.writer(member_file)
                writer.writerow(['mock', 'data'])
                writer.writerow(['line', 'two'])
        response = HTTPResponse(
            body=zip_content.getvalue(),
            status=200,
            headers={"Content-Type": "application/zip"}
        )
        return response
    # All other requests...
    response = HTTPResponse(
```

```
        status=404
    )
    return response
```

This function will only respond to one specifically requested combination of `ownerSlug` and `datasetSlug`. Other combinations will get a 404 response, the status code for a resource that can't be found.

The `io.BytesIO` object is an in-memory buffer that can be processed like a file. It is used by the `zipfile.ZipFile` class to create a ZIP archive. A single member is written to this archive. The member has a header row and a single row of data, making it easy to describe in a Gherkin scenario. The response is built from the bytes in this file, a status code of 200, and a header telling the client the content is a ZIP archive.

This service can be run on the desktop. You can use a browser to interact with this server and confirm it works well enough to test our application.

Now that we've seen the mock service that stands in for Kaggle.com, we can look at how to make the **behave** tool run this service when testing a specific scenario.

Behave fixture

We've added a `fixture.kaggle_server` to the scenario. There are two steps to make this tag start the server process running for a given scenario. These steps are:

1. Define a generator function. This will start a subprocess, yield something, and then kill the subprocess.

2. Define a `before_tag()` function to inject the generator function into the step processing.

Here's a generator function that will update the context, and start the mock Kaggle service.

```python
from collections.abc import Iterator
from typing import Any
import subprocess
```

```python
import time
import os
import sys
from behave import fixture, use_fixture
from behave.runner import Context


@fixture
def kaggle_server(context: Context) -> Iterator[Any]:
    if "environment" not in context:
        context.environment = os.environ
    context.environment["ACQUIRE_BASE_URL"] = "http://127.0.0.1:8080"
    # Save server-side log for debugging
    server = subprocess.Popen(
        [sys.executable, "tests/mock_kaggle_bottle.py"],
    )
    time.sleep(0.5)  # 500 ms delay to allow the service to open a socket
    yield server
    server.kill()
```

The portion of the function before the `yield` statement is used during the scenario setup. This will add value to the context that will be used to start the application under test. After the yielded value has been consumed by the **Behave** runner, the scenario executes. When the scenario is finished, one more value is requested from this generator; this request will execute the statements after the `yield` statement. There's no subsequent `yield` statement; the `StopIteration` is the expected behavior of this function.

This `kaggle_server()` function must be used in a scenario when the `@fixture` tag is present. The following function will do this:

```python
from behave import use_fixture
from behave.runner import Context
```

```
def before_tag(context: Context, tag: str) -> None:
    if tag == "fixture.kaggle_server":
        # This will invoke the definition generator.
        # It consumes a value before and after the tagged scenario.
        use_fixture(kaggle_server, context)
```

When the `@fixture.kaggle_server` tag is present, this function will inject the `kaggle_server()` generator function into the overall flow of processing by the runner. The runner will make appropriate requests of the `kaggle_server()` generator function to start and stop the service.

These two functions are placed into the `environment.py` module where the **behave** tool can find and use them.

Now that we have an acceptance test suite, we can turn to implement the required features of the `acquire` application.

Kaggle access module and refactored main application

The goal, of course, is two-fold:

- Add a `kaggle_client.py` module. The unit tests will confirm this works.

- Rewrite the `acquire.py` module from *Chapter 3, Project 1.1: Data Acquisition Base Application* to add the download feature.

The *Approach* section provides some design guidance for building the application. Additionally, the previous chapter, *Chapter 3, Project 1.1: Data Acquisition Base Application* provides the baseline application into which the new acquisition features should be added.

The acceptance tests will confirm the application works correctly.

Given this extended capability, you are encouraged to hunt for additional, interesting data sets. The new application can be revised and extended to acquire new, interesting data in other formats.

Now that we have data acquired from the web in a tidy, easy-to-use format, we can look at acquiring data that isn't quite so tidy. In the next section, we'll look at how to scrape data out of an HTML page.

Project 1.3: Scrape data from a web page

In some cases, we want data that's provided by a website that doesn't have a tidy API. The data is available via an HTML page. This means the data is surrounded by HTML *markup*, text that describes the semantics or structure of the data.

We'll start with a description of the application, and then move on to talk about the architectural approach. This will be followed with a detailed list of deliverables.

Description

We'll continue to describe projects designed to acquire some data for further analysis. In this case, we'll look at data that is available from a website, but is embedded into the surrounding HTML markup. We'll continue to focus on Anscombe's Quartet data set because it's small and diagnosing problems is relatively simple. A larger data set introduces additional problems with time and storage.

Parts of this application are an extension to the project in *Project 1.2: Acquire data from a web service*. The essential behavior of this application will be similar to the previous project. This project will use a CLI application to grab data from a source.

The User Experience (UX) will also be a command-line application with options to fine-tune the data being gathered. Our expected command line should like something like the following:

```
% python src/acquire.py -o quartet --page
"https://en.wikipedia.org/wiki/Anscombe's_quartet" --caption
"Anscombe's quartet"
```

The `-o quartet` argument specifies a directory into which four results are written. These will have names like `quartet/series_1.json`.

The table is buried in the HTML of the URL given by the --page argument. Within this HTML, the target table has a unique <caption> tag:

<caption>Anscombe's quartet</caption>.

About the source data

This data embedded in HTML markup is generally marked up with the <table> tag. A table will often have the following markup:

```
<table class="wikitable" style="text-align: center; margin-left:auto;
  margin-right:auto;" border="1">
    <caption>Anscombe's quartet</caption>
    <tbody>
      <tr>
        <th colspan="2">I</th>
        etc.
      </tr>
      <tr>
        <td><i>x</i></td>
        <td><i>y</i></td>
        etc.
      </tr>
      <tr>
        <td>10.0</td>
        <td>8.04</td>
        etc.
      </tr>
    </tbody>
</table>
```

In this example, the overall <table> tag will have two child tags, a <caption> and a <tbody>.

The table's body, within `<tbody>`, has a number of rows wrapped in `<tr>` tags. The first row has headings in `<th>` tags. The second row also has headings, but they use the `<td>` tags. The remaining rows have data, also in `<td>` tags.

This structure has a great deal of regularity, making it possible to use a parser like **Beautiful Soup** to locate the content.

The output will match the extraction processing done for the previous projects. See *Chapter 3, Project 1.1: Data Acquisition Base Application*, for the essence of the data acquisition application.

This section has looked at the input and processing for this application. The output will match earlier projects. In the next section, we'll look at the overall architecture of the software.

Approach

We'll take some guidance from the C4 model (`https://c4model.com`) when looking at our approach:

- **Context**: For this project, a context diagram would show a user extracting data from a source. You may find it helpful to draw this diagram.

- **Containers**: One container is the user's personal computer. The other container is the Wikipedia website, which provides the data.

- **Components**: We'll address the components below.

- **Code**: We'll touch on this to provide some suggested directions.

It's important to consider this application as an extension to the project in *Chapter 3, Project 1.1: Data Acquisition Base Application*. The base level of architectural design is provided in that chapter.

In this project, we'll be adding a new `html_extract` module to capture and parse the data. The overall application in the `acquire` module will change to use the new features. The other modules should remain unchanged.

A new architecture that handles the download of HTML data and the extraction of a table from the source data is shown in *Figure 4.4*.

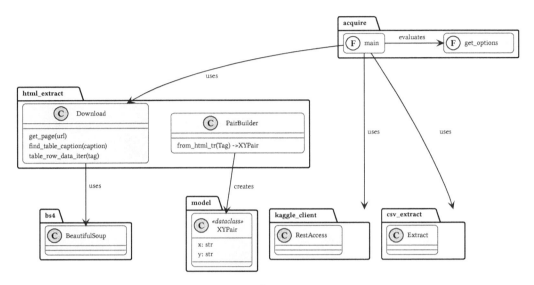

Figure 4.4: Revised Component Design

This diagram suggested classes for the new `html_extract` module. The `Download` class uses `urllib.request` to open the given URL and read the contents. It also uses the `bs4` module (**Beautiful Soup**) to parse the HTML, locate the table with the desired caption, and extract the body of the table.

The `PairBuilder` class hierarchy has four implementations, each appropriate for one of the four data series. Looking back at *Chapter 3, Project 1.1: Data Acquisition Base Application*, there's a profound difference between the table of data shown on the Wikipedia page, and the CSV source file shown in that earlier project. This difference in data organization requires slightly different pair-building functions.

Making an HTML request with urllib.request

The process of reading a web page is directly supported by the `urllib.request` module. The `url_open()` function will perform a GET request for a given URL. The return value is a file-like object — with a `read()` method — that can be used to acquire the content.

This is considerably simpler than making a general RESTful API request where there are a variety of pieces of information to be uploaded and a variety of kinds of results that might be downloaded. When working with common GET requests, this standard library module handles the ordinary processing elegantly.

A suggested design for the first step in the operation is the following function:

```
from urllib.request import urlopen
from bs4 import BeautifulSoup, Tag


def get_page(url: str) -> BeautifulSoup:
    return BeautifulSoup(
        urlopen(url), "html.parser"
    )
```

The `urlopen()` function will open the URL as a file-like object, and provide that file to the `BeautifulSoup` class to parse the resulting HTML.

A `try:` statement to handle potential problems is not shown. There are innumerable potential issues when reaching out to a web service, and trying to parse the resulting content. You are encouraged to add some simple error reporting.

In the next section, we'll look at extracting the relevant table from the parsed HTML.

HTML scraping and Beautiful Soup

The Beautiful Soup data structure has a `find_all()` method to traverse the structure. This will look for tags with specific kinds of properties. This can examine the tag, the attributes, and even the text content of the tag.

See `https://www.crummy.com/software/BeautifulSoup/bs4/doc/#find-all`.

In this case, we need to find a `<table>` tag with a `caption` tag embedded within it. That `caption` tag must have the desired text. This search leads to a bit more complex investigation of the structure. The following function can locate the desired table.

```
def find_table_caption(
```

```
    soup: BeautifulSoup,
    caption_text: str = "Anscombe's quartet"
) -> Tag:
for table in soup.find_all('table'):
    if table.caption:
        if table.caption.text.strip() == caption_text.strip():
            return table
raise RuntimeError(f"<table> with caption {caption_text!r} not found")
```

Some of the tables lack captions. This means the expression `table.caption.text` won't work for string comparison because it may have a `None` value for `table.caption`. This leads to a nested cascade of `if` statements to be sure there's a `<caption>` tag before checking the text value of the tag.

The `strip()` functions are used to remove leading and trailing whitespace from the text because blocks of text in HTML can be surrounded by whitespace that's not displayed, making it surprising when it surfaces as part of the content. Stripping the leading and trailing whitespace makes it easier to match.

The rest of the processing is left for you to design. This processing involves finding all of the `<tr>` tags, representing rows of the table. Within each row (except the first) there will be a sequence of `<td>` tags representing the cell values within the row.

Once the text has been extracted, it's very similar to the results from a `csv.reader`.

After considering the technical approach, it's time to look at the deliverables for this project.

Deliverables

This project has the following deliverables:

- Documentation in the `docs` folder.

- Acceptance tests in the `tests/features` and `tests/steps` folders.

- Unit tests for the application modules in the `tests` folder.

- Mock HTML pages for unit testing will be part of the unit tests.

- Application to acquire data from an HTML page.

We'll look at a few of these deliverables in a little more detail.

Unit test for the html_extract module

The `urlopen()` function supports the `http:` and `https:` schemes. It also supports the `file:` protocol. This allows a test case to use a URL of the form `file:///path/to/a/file.html` to read a local HTML file. This facilitates testing by avoiding the complications of accessing data over the internet.

For testing, it makes sense to prepare files with the expected HTML structure, as well as invalid structures. With some local files as examples, a developer can run test cases quickly.

Generally, it's considered a best practice to mock the `BeautifulSoup` class. A fixture would respond to the various `find_all()` requests with mock tag objects.

When working with HTML, however, it seems better to provide mock HTML. The wide variety of HTML seen in the wild suggests that time spent with real HTML is immensely valuable for debugging.

Creating `BeautifulSoup` objects means the unit testing is more like integration testing. The benefits of being able to test a wide variety of odd and unusual HTML seems to be more valuable than the cost of breaking the ideal context for a unit test.

Having example HTML files plays well with the way **pytest** fixtures work. A fixture can create a file and return the path to the file in the form of a URL. After the test, the fixture can remove the file.

A fixture with a test HTML page might look like this:

```
from pytest import fixture
```

```python
from textwrap import dedent


@fixture
def example_1(tmp_path):
    html_page = tmp_path / "works.html"
    html_page.write_text(
        dedent("""\
        <!DOCTYPE html>
        <html>
        etc.
        </html>
        """
        )
    )
    yield f"file://{str(html_page)}"
    html_page.unlink()
```

This fixture uses the `tmp_path` fixture to provide access to a temporary directory used only for this test. The file, `works.html`, is created, and filled with an HTML page. The test case should include multiple `<table>` tags, only one of which was the expected `<caption>` tag.

The `dedent()` function is a handy way to provide a long string that matches the prevailing Python indent. The function removes the indenting whitespace from each line; the resulting text object is not indented.

The return value from this fixture is a URL that can be used by the `urlopen()` function to open and read this file. After the test is completed, the final step (after the `yield` statement) will remove the file.

A test case might look something like the following:

```python
def test_steps(example_1):
    soup = html_extract.get_page(example_1)
```

```
table_tag = html_extract.find_table_caption(soup, "Anscombe's quartet")
rows = list(html_extract.table_row_data_iter(table_tag))
assert rows == [
    [],
    ['Keep this', 'Data'],
    ['And this', 'Data'],
]
```

The test case uses the `example_1` fixture to create a file and return a URL referring to the file. The URL is provided to a function being tested. The functions within the `html_extract` module are used to parse the HTML, locate the target table, and extract the individual rows.

The return value tells us the functions work properly together to locate and extract data. You are encouraged to work out the necessary HTML for good — and bad — examples.

Acceptance tests

As noted above in *Unit test for the html_extract module*, the acceptance test case HTML pages can be local files. A scenario can provide a local `file://` URL to the application and confirm the output includes properly parsed data.

The Gherkin language permits including large blocks of text as part of a scenario.

We can imagine writing the following kinds of scenarios in a feature file:

```
Scenario: Finds captioned table and extracts data
  Given an HTML page "example_1.html"
    """
      <!DOCTYPE html>
      <html>
        etc. with multiple tables.
      </html>
    """
  When we run the html extract command
```

```
Then log has INFO line with "header: ['Keep this', 'Data']"
And log has INFO line with "count: 1"
```

The HTML extract command is quite long. The content is available as the `context.text` parameter of the step definition function. Here's what the step definition for this given step looks like:

```python
from textwrap import dedent


@given(u'an HTML page "{filename}"')
def step_impl(context, filename):
    context.path = Path(filename)
    context.path.write_text(dedent(context.text))
    context.add_cleanup(context.path.unlink)
```

The step definition puts the path into the context and then writes the HTML page to the given path. The `dedent()` function removes any leading spaces that may have been left in place by the **behave** tool. Since the path information is available in the context, it can be used by the **When** step. The `context.add_cleanup()` function will add a function that can be used to clean up the file when the scenario is finished. An alternative is to use the environment module's `after_scenario()` function to clean up.

This scenario requires an actual path name for the supplied HTML page to be injected into the text. For this to work out well, the step definition needs to build a command from pieces. Here's one approach:

```python
@when(u'we run the html extract command')
def step_impl(context):
    command = [
        'python', 'src/acquire.py',
        '-o', 'quartet',
        '--page', '$URL',
        '--caption', "Anscombe's quartet"
```

```
    ]
url = f"file://{str(context.path.absolute())}"
command[command.index('$URL')] = url
print(shlex.join(command))
# etc. with subprocess.run() to execute the command
```

In this example, the command is broken down into individual parameter strings. One of the strings must be replaced with the actual file name. This works out nicely because the subprocess.run() function works well with a parsed shell command. The shlex.split() function can be used to decompose a line, honoring the complex quoting rules of the shell, into individual parameter strings.

Now that we have an acceptance test suite, we may find the acquire application doesn't pass all of the tests. It's helpful to define done via an acceptance test and then develop the required HTML extract module and refactor the main application. We'll look at these two components next.

HTML extract module and refactored main application

The goal for this project is two-fold:

- Add an html_extract.py module. The unit tests will confirm this module works.

- Rewrite the acquire.py module from *Chapter 3, Project 1.1: Data Acquisition Base Application* to add the HTML download and extract the feature.

The *Approach* section provides some design guidance for building the application. Additionally, the previous chapter, *Chapter 3, Project 1.1: Data Acquisition Base Application*, provides the baseline application into which the new acquisition features should be added.

The acceptance tests will confirm the application works correctly to gather data from the Kaggle API.

Given this extended capability, you can hunt for data sets that are presented in web pages. Because of the consistency of Wikipedia, it is a good source of data. Many other sites provide relatively consistent HTML tables with interesting data.

In these two projects, we've extended our ability to acquire data from a wide variety of sources.

Summary

This chapter's projects have shown examples of the following features of a data acquisition application:

- Web API integration via the **requests** package. We've used the Kaggle API as an example of a RESTful API that provides data for download and analysis.

- Parsing an HTML web page using the **Beautiful Soup** package.

- Adding features to an existing application and extending the test suite to cover these new alternative data sources.

A challenging part of both of these projects is creating a suite of acceptance tests to describe the proper behavior. Pragmatically, a program without automated tests cannot be trusted. The tests are every bit as important as the code they're exercising.

In some enterprises, the definition of done is breezy and informal. There may be a presentation or an internal memo or a whitepaper that describes the desired software. Formalizing these concepts into tangible test cases is often a significant effort. Achieving agreements can become a source of turmoil as stakeholders slowly refine their understanding of how the software will behave.

Creating mock web services is fraught with difficulty. Some API's permit downloading an `openapi.json` file with the definition of the API complete with examples. Having concrete examples, provided by the host of the API, makes it much easier to create a mock service. A mock server can load the JSON specification, navigate to the example, and provide the official response.

Lacking an OpenAPI specification with examples, developers need to write spike solutions that download detailed responses. These responses can then be used to build mock objects. You are strongly encouraged to write side-bar applications to explore the Kaggle API to see how it works.

In the next chapter, we'll continue this data extraction journey to include extracting data from SQL databases. Once we've acquired data, we'll want to inspect it. *Chapter 6, Project 2.1: Data Inspection Notebook*, will introduce an inspection step.

Extras

Here are some ideas for you to add to these projects.

Locate more JSON-format data

A search of Kaggle will turn up some other interesting data sets in JSON format.

- `https://www.kaggle.com/datasets/rtatman/iris-dataset-json-version`: This data set is famous and available in a number of distinct formats.

- `https://www.kaggle.com/datasets/conoor/stack-overflow-tags-usage`

- `https://www.kaggle.com/datasets/queyrusi/the-warship-dataset`

One of these is a JSON download. The other two are ZIP archives that contain JSON-format content.

This will require revising the application's architecture to extract the JSON format data instead of CSV format data.

An interesting complication here is the distinction between CSV data and JSON data:

- CSV data is pure text, and later conversions are required to make useful Python objects.

- Some JSON data is converted to Python objects by the parser. Some data (like datestamps) will be left as text.

At acquisition time, this doesn't have a significant impact. However, when we get to *Chapter 9, Project 3.1: Data Cleaning Base Application*, we'll have to account for data in a text-only form distinct from data with some conversions applied.

The Iris data set is quite famous. You can expand on the designs in this chapter to acquire Iris data from a variety of sources. The following steps could be followed:

1. Start with the Kaggle data set in JSON format. Build the needed model, and extract modules to work with this format.

2. Locate other versions of this data set in other formats. Build the needed extract modules to work with these alternative formats.

Once a core acquisition project is complete, you can leverage this other famous data set as an implementation choice for later projects.

Other data sets to extract

See the **CO$_2$ PPM — Trends in Atmospheric Carbon Dioxide** data set, available at `https://datahub.io/core/co2-ppm`, for some data that is somewhat larger. This page has a link to an HTML table with the data.

See `https://datahub.io/core/co2-ppm/r/0.html` for a page with the complete data set as an HTML table. This data set is larger and more complicated than Anscombe's Quartet. In *Chapter 6, Project 2.1: Data Inspection Notebook*, we'll address some of the special cases in this data set.

Handling schema variations

The two projects in this chapter each reflect a distinct schema for the source data.

One CSV format can be depicted via an **Entity-Relationship Diagram (ERD)**, shown in *Figure 4.5*.

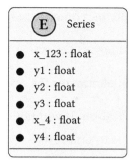

Figure 4.5: Source entity-relationship diagram

One column, x_123, is the x-value of three distinct series. Another column, x_4, is the x-value for one series.

A depiction of the HTML format as an ERD is shown in *Figure 4.6.*

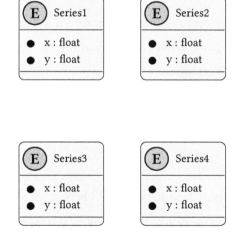

Figure 4.6: Notional entity-relationship diagram

The x-values are repeated as needed.

This difference requires several distinct approaches to extracting the source data.

In these projects, we've implemented this distinction as distinct subclasses of a `PairBuilder` superclass.

An alternative design is to create distinct functions with a common type signature:

```
PairBuilder: TypeVar = Callable[[list[str]], XYPair]
```

Making each conversion a function eliminates the overhead of the class definition.

This rewrite can be a large simplification. It will not change any acceptance tests. It will, however, require numerous changes to unit tests.

The functional design offers some simplification over class-based design. You are encouraged to perform the functional redesign of the suggestions in this book.

CLI enhancements

The CLI for these two projects is left wide open, permitting a great deal of design flexibility and alternatives. Because the CLI is part of externally visible behavior, it becomes necessary to write acceptance tests for the various CLI options and arguments.

As noted in *Additional acceptance scenarios*, there are a number of acceptance test scenarios that are not on the "happy path" where the application works. These additional scenarios serve to catalog a number of erroneous uses of the application.

This becomes more important as more features are added and the CLI becomes more complicated. You are encouraged to write acceptance tests for invalid CLI use.

Logging

Logging is an important part of data acquisition. There are a number of potential problems exposed by these two projects. A website might be unresponsive, or the API may have changed. The HTML may have been reformatted in some subtle way.

A *debug* or *verbose* mode should be available to expose the interactions with external services to be sure of the HTTP status codes and headers.

Additionally, count values should be displayed to summarize the bytes downloaded, the lines of text examined, and the number of XYPair objects created. The idea is to characterize the inputs, the various processing steps, and the outputs.

These counts are essential for confirming that data is processed and filtered correctly. They're an important tool for making parts of the processing more observable. A user wants to confirm that all of the downloaded data is either part of the results or filtered and discarded for a good reason.

You are encouraged to include counts for input, processing, and output in the log.

5

Data Acquisition Features: SQL Database

In this chapter, you will be guided through two projects that demonstrate how to work with SQL databases as a source of data for analysis. This will build on the foundational application built in the previous two chapters.

This chapter will focus on SQL extracts. Since enterprise SQL databases tend to be very private, we'll guide the reader through creating an SQLite database first. This database will be a stand-in for a private enterprise database. Once there's a database available, we will look at extracting data from the database.

This chapter's projects cover the following essential skills:

- Building SQL databases.

- Extracting data from SQL databases.

The first project will build a SQL database for use by the second project.

In an enterprise environment, the source databases will already exist.

On our own personal computers, these databases don't exist. For this reason, we'll build a database in the first project, and extract from the database in the second project.

We'll start by looking at getting data into a SQL database. This will be a very small and simple database; the project will steer clear of the numerous sophisticated design complications for SQL data.

The second project will use SQL queries to extract data from the database. The objective is to produce data that is consistent with the projects in the previous chapters.

Project 1.4: A local SQL database

We'll often need data stored in a database that's accessed via the SQL query language. Use a search string like "SQL is the lingua franca" to find numerous articles offering more insight into the ubiquity of SQL. This seems to be one of the primary ways to acquire enterprise data for further analysis.

In the previous chapter, *Chapter 4, Data Acquisition Features: Web APIs and Scraping*, the projects acquired data from publicly available APIs and web pages. There aren't many publicly available SQL data sources. In many cases, there are dumps (or exports) of SQLite databases that can be used to build a local copy of the database. Direct access to a remote SQL database is not widely available. Rather than try to find access to a remote SQL database, it's simpler to create a local SQL database. The SQLite database is provided with Python as part of the standard library, making it an easy choice.

You may want to examine other databases and compare their features with SQLite. While some databases offer numerous capabilities, doing SQL extracts rarely seems to rely on anything more sophisticated than a basic SELECT statement. Using another database may require some changes to reflect that database's connections and SQL statement execution. For the most part, the DB-API interface in Python is widely used; there may be unique features for databases other than SQLite.

We'll start with a project to populate the database. Once a database is available, you can then move on to a more interesting project to extract the data using SQL statements.

Description

The first project for this chapter will prepare a SQL database with data to analyze. This is a necessary preparation step for readers working outside an enterprise environment with accessible SQL databases.

One of the most fun small data sets to work with is Anscombe's Quartet.

`https://www.kaggle.com/datasets/carlmcbrideellis/data-anscombes-quartet`

The URL given above presents a page with information about the CSV format file. Clicking the **Download** button will download the small file of data to your local computer.

The data is available in this book's GitHub repository's `data` folder, also.

In order to load a database, the first step is designing the database. We'll start with a look at some table definitions.

Database design

A SQL database is organized as tables of data. Each table has a fixed set of columns, defined as part of the overall database schema. A table can have an indefinite number of rows of data.

For more information on SQL databases, see `https://www.packtpub.com/product/learn-sql-database-programming/9781838984762` and `https://courses.packtpub.com/courses/sql`.

Anscombe's Quartet consists of four series of (x, y) pairs. In one commonly used source file, three of the series share common x values, whereas the fourth series has distinct x values.

A relational database often decomposes complicated entities into a collection of simpler entities. The objective is to minimize the repetitions of association types. The Anscombe's Quartet information has four distinct series of data values, which can be represented as the following two types of entities:

- The series is composed of a number of individual values. A table named `series_value` can store the individual values that are part of a series.

- A separate entity has identifying information for the series as a whole. A table named `sources` can store identifying information.

This design requires the introduction of key values to uniquely identify the series, and connect each value of a series with the summary information for the series.

 For Anscombe's Quartet data, the summary information for a series is little more than a name.

This design pattern of an overall summary and supporting details is so common that it is essential for this project to reflect that common pattern.

See *Figure 5.1* for an ERD that shows the two tables that implement these entities and their relationships.

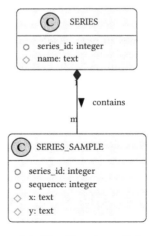

Figure 5.1: The Database Schema

This project will create a small application to build this schema of two tables. This application will can then load data into these tables.

Data loading

The process of loading data involves three separate operations:

- Reading the source data from a CSV (or other format) file.

- Executing SQL INSERT statements to create rows in tables.

- Executing a COMMIT to finalize the transaction and write data to the underlying database files.

Prior to any of these steps, the schema must be defined using CREATE TABLE statements.

In a practical application, it's also common to offer a composite operation to drop the tables, recreate the schema, and then load the data. The rebuilding often happens when exploring or experimenting with database designs. Many times, an initial design will prove unsatisfactory, and changes are needed. Additionally, the idea of building (and rebuilding) a small database will also be part of the acceptance test for any data acquisition application.

In the next section, we'll look at how to create a SQL database that can serve as a surrogate for a production database in a large enterprise.

Approach

There are two general approaches to working with SQL databases for this kind of test or demonstration application:

- Create a small application to build and populate a database.

- Create a SQL script via text formatting and run this through the database's CLI application. See https://sqlite.org/cli.html.

The small application will make use of the database client connection to execute SQL statements. In this case, a single, generic INSERT statement template with placeholders can be used. The client connection can provide values for the placeholders. While the application isn't complex, it will require unit and acceptance test cases.

The SQL script alternative uses a small application to transform data rows into valid INSERT statements. In many cases, a text editor search-and-replace can transform data text into INSERT statements. For more complex cases, Python f-strings can be used. The f-string might look like the following:

```
print(
    f"INSERT INTO SSAMPLES(SERIES, SEQUENCE, X, Y)"
    f"VALUES({series}, {sequence}, '{x}', '{y}')"
)
```

This is often successful but suffers from a potentially severe problem: a *SQL injection exploit.*

The SQL injection exploit works by including an end-of-string-literal apostrophe ' in a data value. This can lead to an invalid SQL statement. In extreme cases, it can allow injecting additional SQL statements to transform the INSERT statement into a script. For more information, see https://owasp.org/www-community/attacks/SQL_Injection. Also, see https://xkcd.com/327/ for another example of a SQL injection exploit.

While SQL injection can be used maliciously, it can also be a distressingly common accident. If a text data value happens to have ' in it, then this can create a statement in the SQL script file that has invalid syntax. SQL cleansing only defers the problem to the potentially complicated SQL cleansing function.

It's simpler to avoid building SQL text in the first place. A small application can be free from the complications of building SQL text.

We'll start by looking at the data definition for this small schema. Then we'll look at the data manipulation statements. This will set the stage for designing the small application to build the schema and load the data.

SQL Data Definitions

The essential data definition in SQL is a table with a number of columns (also called *attributes*). This is defined by a CREATE TABLE statement. The list of columns is provided in this statement. In addition to the columns, the language permits table constraints to further refine how a table may be used. For our purposes, the two tables can be defined as follows:

```
CREATE TABLE IF NOT EXISTS series(
    series_id INTEGER,
    name TEXT,

    PRIMARY KEY (series_id)
);


CREATE TABLE IF NOT EXISTS series_sample(
    series_id INTEGER,
    sequence INTEGER,
    x TEXT,
    y TEXT,

    PRIMARY KEY (series_id, sequence),
    FOREIGN KEY (series_id) REFERENCES series(series_id)
);
```

To remove a schema, the `DROP TABLE IF EXISTS series_sample` and `DROP TABLE IF EXISTS series` statements will do what's needed. Because of the foreign key reference, some databases make it necessary to remove all of the related `series_sample` rows before a `series` row can be removed.

The `IF EXISTS` and `IF NOT EXISTS` clauses are handy when debugging. We may, for example, change the SQL and introduce a syntax error into one of the `CREATE TABLE` statements. This can leave an incomplete schema. After fixing the problem, simply rerunning the entire sequence of `CREATE TABLE` statements will create only the tables that were missing.

An essential feature of this example SQL data model is a simplification of the data types involved. Two columns of data in the `series_sample` table are both defined as TEXT. This is a rarity; most SQL databases will use one of the available numeric types.

While SQL data has a variety of useful types, the raw data from other applications, however, isn't numeric. CSV files and HTML pages only provide text. For this reason, the results from this application need to be text, also. Once the tables are defined, an application can insert rows.

SQL Data Manipulations

New rows are created with the INSERT statement. While SQLite allows some details to be omitted, we'll stick with a slightly wordier but more explicit statement. Rows are created in the two tables as follows:

```
INSERT INTO series(series_id, name) VALUES(:series_id, :name)


INSERT INTO series_sample(series_id, sequence, x, y)
  VALUES(:series_id, :sequence, :x, :y)
```

The identifiers with a colon prefix, :x, :y, :series_id, etc., are parameters that will be replaced when the statement is executed. Since these replacements don't rely on SQL text rules — like the use of apostrophes to end a string — any value can be used.

It's rare to need to delete rows from these tables. It's easier (and sometimes faster) to drop and recreate the tables when replacing the data.

SQL Execution

Python's SQLite interface is the `sqlite3` module. This conforms to the PEP-249 standard (`https://peps.python.org/pep-0249/`) for database access. An application will create a database connection in general. It will use the connection to create a *cursor*, which can query or update the database.

The connection is made with a connection string. For many databases, the connection string will include the server hosting the database, and the database name; it may also include security credentials or other options. For SQLite, the connection string can be a complete URI with the form `file:filename.db`. This has a scheme, `file:` and a path to the database file.

It's not required by this application, but a common practice is to sequester the SQL statements into a configuration file. Using a TOML format can be a handy way to separate the processing from the SQL statements that implement the processing. This separation permits small SQL changes without having to change the source files. For compiled languages, this is essential. For Python, it's a helpful way to make SQL easier to find when making database changes.

A function to create the schema might look like this:

```
CREATE_SERIES = """
CREATE TABLE IF NOT EXISTS series(
-- rest of the SQL shown above...
"""


CREATE_VALUES = """
CREATE TABLE IF NOT EXISTS series_sample(
-- rest of the SQL shown above...
"""


CREATE_SCHEMA = [
    CREATE_SERIES,
    CREATE_VALUES
]


def execute_statements(
        connection: sqlite3.Connection,
        statements: list[str]
) -> None:
```

```
for statement in statements:
    connection.execute(statement)
connection.commit()
```

The `CREATE_SCHEMA` is the sequence of statements required to build the schema. A similar sequence of statements can be defined to drop the schema. The two sequences can be combined to drop and recreate the schema as part of ordinary database design and experimentation.

A main program can create the database with code similar to the following:

```
with sqlite3.connect("file:example.db", uri=True) as connection:
    schema_build_load(connection, config, data_path)
```

This requires a function, `schema_build_load()`, to drop and recreate the schema and then load the individual rows of data.

We'll turn to the next step, loading the data. This begins with loading the series definitions, then follows this with populating the data values for each series.

Loading the SERIES table

The values in the `SERIES` table are essentially fixed. There are four rows to define the four series.

Executing a SQL data manipulation statement requires two things: the statement and a dictionary of values for the placeholders in the statement.

In the following code sample, we'll define the statement, as well as four dictionaries with values for placeholders:

```
INSERT_SERIES = """
    INSERT INTO series(series_id, name)
        VALUES(:series_id, :name)
"""
```

```
SERIES_ROWS = [
    {"series_id": 1, "name": "Series I"},
    {"series_id": 2, "name": "Series II"},
    {"series_id": 3, "name": "Series III"},
    {"series_id": 4, "name": "Series IV"},
]


def load_series(connection: sqlite3.Connection) -> None:
    for series in SERIES_ROWS:
        connection.execute(INSERT_SERIES, series)
    connection.commit()
```

The execute() method of a connection object is given the SQL statement with placeholders and a dictionary of values to use for the placeholders. The SQL template and the values are provided to the database to insert rows into the table.

For the individual data values, however, something more is required. In the next section, we'll look at a transformation from source CSV data into a dictionary of parameter values for a SQL statement.

Loading the SERIES_VALUE table

It can help to refer back to the project in *Chapter 3, Project 1.1: Data Acquisition Base Application*. In this chapter, we defined a dataclass for the (*x*, *y*) pairs, and called it XYPair. We also defined a class hierarchy of PairBuilder to create XYPair objects from the CSV row objects.

It can be confusing to load data using application software that is suspiciously similar to the software for extracting data.

This confusion often arises in cases like this where we're forced to build a demonstration database.

It can also arise in cases where a test database is needed for complex analytic applications.

 In most enterprise environments, the databases already exist and are already full of data. Test databases are still needed to confirm that analytic applications work.

The INSERT statement, shown above in *SQL Data Manipulations* has four placeholders. This means a dictionary with four parameters is required by the execute() method of a connection.

The dataclasses module includes a function, asdict(), to transform the object of the XYPair into a dictionary. This has two of the parameters required, :x and :y.

We can use the | operator to merge two dictionaries together. One dictionary has the essential attributes of the object, created by asdict(). The other dictionary is the SQL overheads, including a value for :series_id, and a value for :sequence.

Here's a fragment of code that shows how this might work:

```python
for sequence, row in enumerate(reader):
    for series_id, extractor in SERIES_BUILDERS:
        param_values = (
            asdict(extractor(row)) |
            {"series_id": series_id, "sequence": sequence}
        )
        connection.execute(insert_values_SQL, param_values)
```

The reader object is a csv.DictReader for the source CSV data. The SERIES_BUILDERS object is a sequence of two-tuples with the series number and a function (or callable object) to extract the appropriate columns and build an instance of XYPair.

For completeness, here's the value of the SERIES_BUILDERS object:

```python
SERIES_BUILDERS = [
```

```
    (1, series_1),
    (2, series_2),
    (3, series_3),
    (4, series_4)
]
```

In this case, individual functions have been defined to extract the required columns from the CSV source dictionary and build an instance of XYPair.

The above code snippets need to be built as proper functions and used by an overall main() function to drop the schema, build the schema, insert the values for the SERIES table, and then insert the SERIES_VALUE rows.

A helpful final step is a query to confirm the data was loaded. Consider something like this:

```
SELECT s.name, COUNT(*)
  FROM series s JOIN series_sample sv
    ON s.series_id = sv.series_id
  GROUP BY s.series_id
```

This should report the names of the four series and the presence of 11 rows of data.

Deliverables

There are two deliverables for this mini-project:

- A database for use in the next project. The primary goal is to create a database that is a surrogate for a production database in use by an enterprise.

- An application that can build (and rebuild) this database. This secondary goal is the means to achieve the primary goal.

Additionally, of course, unit tests are strongly encouraged. This works out well when the application is designed for testability. This means two features are essential:

- The database connection object is created in the main() function.

- The connection object is passed as an argument value to all the other functions that
  interact with the database.

Providing the connection as a parameter value makes it possible to test the various functions
isolated from the overhead of a database connection. The tests for each application function
that interacts with the database are given a mock connection object. Most mock connection
objects have a mock `execute()` method, which returns a mock cursor with no rows. For
queries, the mock `execute()` method can return mocked data rows, often something as
simple as a `sentinel` object.

After exercising a function, the mock `execute()` method can then be examined to be sure
the statement and parameters were provided to the database by the application.

A formal acceptance test for this kind of one-use-only application seems excessive. It seems
easier to run the application and look at the results with a SQL `SELECT` query. Since the
application drops and recreates the schema, it can be re-run until the results are acceptable.

Project 1.5: Acquire data from a SQL extract

At this point, you now have a useful SQL database with schema and data. The next step is
to write applications to extract data from this database into a useful format.

Description

It can be difficult to use an operational database for analytic processing. During normal
operations, locking is used to assure that database changes don't conflict with or overwrite
each other. This locking can interfere with gathering data from the database for analytic
purposes.

There are a number of strategies for extracting data from an operational database. One
technique is to make a backup of the operational database and restore it into a temporary
clone database for analytic purposes. Another technique is to use any replication features
and do analytical work in the replicated database.

The strategy we'll pursue here is the "table-scan" approach. It's often possible to do rapid queries without taking out any database locks. The data may be inconsistent because of in-process transactions taking place at the time the query was running. In most cases, the number of inconsistent entities is a tiny fraction of the available data.

If it's necessary to have a *complete and consistent* snapshot at a specific point in time, the applications need to have been designed with this idea in mind. It can be very difficult to establish the state of a busy database with updates being performed by poorly designed applications. In some cases, the definitions of *complete* and *consistent* may be difficult to articulate because the domain of state changes isn't known in enough detail.

It can be frustrating to work with poorly designed databases.

 It's often important to educate potential users of analytic software on the complexities of acquiring the data. This education needs to translate the database complications into the effect on the decisions they're trying to make and the data that supports those decisions.

The **User Experience (UX)** will be a command-line application. Our expected command line should look something like the following:

```
% python src/acquire.py -o quartet --schema extract.toml \
    --db_uri file:example.db -u username

Enter your password:
```

The -o quartet argument specifies a directory into which four results are written. These will have names like quartet/series_1.json.

The --schema extract.toml argument is the name of a file with the SQL statements that form the basis for the database queries. These are kept separate from the application to make it slightly easier to respond to the database structure changes without rewriting the application program.

The --db_uri file:example.db argument provides the URI for the database. For SQLite, the URIs have a scheme of `file:` and a path to the database file. For other database engines, the URI may be more complicated.

The `-u` argument provides a username for connecting to the database. The password is requested by an interactive prompt. This keeps the password hidden.

 The UX shown above includes a username and password.

While it won't actually be needed for SQLite, it will be needed for other databases.

The Object-Relational Mapping (ORM) problem

A relational database design decomposes complicated data structures into a number of simpler entity types, which are represented as tables. The process of decomposing a data structure into entities is called *normalization*. Many database designs fit a pattern called *Third Normal Form*; but there are additional normalization forms. Additionally, there are compelling reasons to break some of the normalization rules to improve performance.

The relational normalization leads to a consistent representation of data via simple tables and columns. Each column will have an atomic value that cannot be further decomposed. Data of arbitrary complexity can be represented in related collections of flat, normalized tables.

See `https://www.packtpub.com/product/basic-relational-database-design-video /9781838557201` for some more insights into the database design activity.

The process of retrieving a complex structure is done via a relational *join* operation. Rows from different tables and joined into a result set from which Plain Old Python Objects can be constructed. This join operation is part of the `SELECT` statement. It appears in the `FROM` clause as a rule that states how to match rows in one table with rows from another table.

This distinction between relational design and object-oriented design is sometimes called the *Object-Relational Impedance Mismatch*. For more background, see `https://wiki.c2. com/?ObjectRelationalImpedanceMismatch`.

One general approach to reading complex data from a relational database is to create to an ORM layer. This layer uses SQL SELECT statements to extract data from multiple tables to build a useful object instance. The ORM layer may use a separate package, or it may be part of the application. While an ORM design can be designed poorly — i.e. the ORM-related operations may be scattered around haphazardly — the layer is always present in any application.

There are many packages in the **Python Package Index(PyPI)** that offer elegant, generalized ORM solutions. The **SQLAlchemy** (`https://www.sqlalchemy.org`) package is very popular. This provides a comprehensive approach to the entire suite of **Create, Retrieve, Update, and Delete(CRUD)** operations.

There are two conditions that suggest creating the ORM layer manually:

- Read-only access to a database. A full ORM will include features for operations that won't be used.

- An oddly designed schema. It can sometimes be difficult to work out an ORM definition for an existing schema with a design that doesn't fit the ORM's built-in assumptions.

There's a fine line between a bad database design and a confusing database design. A bad design has quirky features that cannot be successfully described through an ORM layer. A confusing design can be described, but it may require using "advanced" features of the ORM package. In many cases, building the ORM mapping requires learning enough about the ORM's capabilities to see the difference between bad and confusing.

In many cases, a relational schema may involve a vast number of interrelated tables, sometimes from a wide variety of subject areas. For example, there may be products and a product catalog, sales records for products, and inventory information about products. What is the proper boundary for a "product" class? Should it include everything in the database related to a product? Or should it be limited by some bounded context or problem domain?

Considerations of existing databases should lead to extensive conversations with users on the problem domain and context. It also leads to further conversations with the owners of the applications creating the data. All of the conversations are aimed at understanding how a user's concept may overlap with existing data sources.

Acquiring data from relational databases can be a challenge.

The relational normalization will lead to complications. The presence of overlapping contexts can lead to further complications.

What seems to be helpful is providing a clear translation from the technical world of the database to the kinds of information and decisions users want to make.

About the source data

See *Figure 5.2* for an ERD that shows the two tables that provide the desired entities:

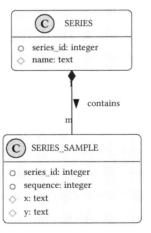

Figure 5.2: The Database Schema

In the design shown above, two tables decompose instances of the `Series` class. Here are the Python class definitions:

```
from dataclasses import dataclass
```

```
@dataclass
class SeriesSample:
    x: str
    y: str


@dataclass
class Series:
    name: str
    samples: list[SeriesSample]
```

The idea here is that a collection of `SeriesSample` objects are part of a single composite `Series` object. The `SeriesSample` objects, separated from the containing `Series`, aren't useful in isolation. A number of `SeriesSample` instances depend on a `Series` object.

There are three general approaches to retrieving information from a normalized collection of tables:

- A single SQL query. This forces the database server to **join** rows from multiple tables, providing a single result set.

- A series of queries to extract data from separate tables and then do lookups using Python dictionaries.

- Nested SQL queries. These use simpler SQL but can make for a large number of database requests.

Neither alternative is a perfect solution in all cases. Many database designers will insist that database join operations are magically the fastest. Some actual timing information suggests that Python dictionary lookups can be much faster. Numerous factors impact query performance and the prudent design is to implement alternatives and compare performance.

The number of factors influencing performance is large. No simple "best practice" exists. Only actual measurements can help to make a design decision.

The join query to retrieve the data might look this:

```
SELECT s.name, sv.x, sv.y
  FROM series s JOIN series_sample sv ON s.series_id = sv.series_id
```

Each distinct value of s.name will lead to the creation of a distinct Series object. Each row of sv.x, and sv.y values becomes a SeriesSample instance within the Series object.

Building objects with two separate SELECT statements involves two simpler queries. Here's the "outer loop" query to get the individual series:

```
SELECT s.name, s.series_id
  FROM series s
```

Here's the "inner loop" query to get rows from a specific series:

```
SELECT sv.x, sv.y
  FROM series_sample sv
  WHERE sv.series_id = :series_id
  ORDER BY sv.sequence
```

The second SELECT statement has a placeholder that depends on the results of the first query. The application must provide this parameter when making a nested request for a series-specific subset of rows from the series_sample table.

It's also important to note the output is expected to be pure text, which will be saved in ND JSON files. This means the sophisticated structure of the SQL database will be erased.

This will also make the interim results consistent with CSV files and HTML pages, where the data is only text. The output should be similar to the output from the CSV extract in *Chapter 3, Project 1.1: Data Acquisition Base Application*: a file of small JSON documents that have the keys "x" and "y". The goal is to strip away structure that may have been imposed by the data persistence mechanism — a SQL database for this project. The data is reduced into a common base of text.

In the next section, we'll look more closely at the technical approach to acquiring data from a SQL database.

Approach

We'll take some guidance from the C4 model (`https://c4model.com`) when looking at our approach.

- **Context**: For this project, a context diagram would show a user extracting data from a source. The reader may find it helpful to draw this diagram.

- **Containers**: One container is the user's personal computer. The other container is the database server, which is running on the same computer.

- **Components**: We'll address the components below.

- **Code**: We'll touch on this to provide some suggested directions.

This project adds a new `db_client` module to extract the data from a database. The overall application in the `acquire` module will change to make use of this new module. The other modules — for the most part — will remain unchanged.

The component diagram in *Figure 5.3* shows an approach to this project.

This diagram shows a revision to the underlying `model`. This diagram extends the `model` module to make the distinction between the composite series object and the individual samples within the overall series. It also renames the old `XYPair` class to a more informative `SeriesSample` class.

This distinction between series has been an implicit part of the project in the previous chapters. At this point, it seems potentially helpful to distinguish a collection of samples from an individual sample.

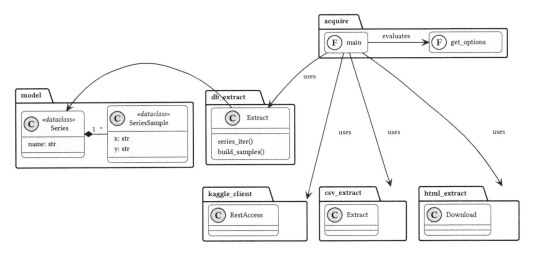

Figure 5.3: Component Diagram

Some readers may object to renaming a class partway through a series of closely related projects. This kind of change is — in the author's experience — very common. We start with an understanding that evolves and grows the more we work the problem domain, the users, and the technology. It's very difficult to pick a great name for a concept. It's more prudent to fix names as we learn.

The new module will make use of two SQL queries to perform the extract. We'll look at these nested requests in the next section.

Extract from a SQL DB

The extraction from the database constructs a series of two parts. The first part is to get the attributes of the Series class. The second part is to get each of the individual SeriesSample instances.

Here's the overview of a potential class design:

```python
import model
import sqlite3
from typing import Any
from collections.abc import Iterator
```

```
class Extract:
    def build_samples(
            self,
            connection: sqlite3.Connection,
            config: dict[str, Any],
            name: str
    ) -> model.Series:

        . . .


    def series_iter(
            self,
            connection: sqlite3.Connection,
            config: dict[str, Any]
    ) -> Iterator[model.Series]:

        . . .
```

The series_iter() method iterates over the Series instances that can be created from the database. The build_samples() method creates the individual samples that belong to a series.

Here's a first draft of an implementation of the build_samples() method:

```
def build_samples(
        self,
        connection: sqlite3.Connection,
        config: dict[str, Any],
        name: str
) -> list[model.SeriesSample]:
    samples_cursor = connection.execute(
        config['query']['samples'],
        {"name": name}
```

```
    )
    samples = [
        model.SeriesSample(
            x=row[0],
            y=row[1])
        for row in samples_cursor
    ]
    return samples
```

This method will extract the collection of samples for a series given the name. It relies on the SQL query in the config object. The list of samples is built from the results of the query using a list comprehension.

> This first draft implementation has a dependency on the SeriesSample class name. This is another SOLID design issue, similar to the one in *Class design* of *Chapter 3, Project 1.1: Data Acquisition Base Application.*
>
> A better implementation would replace this direct dependency with a dependency that can be injected at runtime, permitting better isolation for unit testing.

Here's an implementation of the series_iter() method:

```
def series_iter(
        self,
        connection: sqlite3.Connection,
        config: dict[str, Any]
) -> Iterator[model.Series]:
    print(config['query']['names'])
    names_cursor = connection.execute(config['query']['names'])
    for row in names_cursor:
        name=row[0]
```

```
    yield model.Series(
        name=name,
        samples=self.build_samples(connection, config, name)
    )
```

This method will extract each of the series from the database. It, too, gets the SQL statements from a configuration object, `config`. A configuration object is a dictionary of dictionaries. This structure is common for TOML files.

The idea is to have a configuration file in TOML notation that looks like this:

```
[query]
summary =   """
SELECT s.name, COUNT(*)
  FROM series s JOIN series_sample sv ON s.series_id = sv.series_id
  GROUP BY s.series_id
"""


detail =   """
SELECT s.name, s.series_id, sv.sequence, sv.x, sv.y
  FROM series s JOIN series_value sv ON s.series_id = sv.series_id
"""


names = """
SELECT s.name FROM series s
"""


samples = """
SELECT sv.x, sv.y
  FROM series_sample sv JOIN series s ON s.series_id = sv.series_id
  WHERE s.name = :name
  ORDER BY sv.sequence
```

```
"""
```

This configuration has a [query] section, with several individual SQL statements used to query the database. Because the SQL statements are often quite large, triple quotes are used to delimit them.

In cases where the SQL statements are very large, it's can seem helpful to put them in separate files. This leads to a more complicated configuration with a number of files, each with a separate SQL statement.

Before we look at the deliverables, we'll talk a bit about why this data acquisition application is different from the previous projects.

SQL-related processing distinct from CSV processing

It's helpful to note some important distinctions between working with CSV data and working with SQL data.

First, CSV data is always text. When working with a SQL database, the underlying data often has a data type that maps pleasantly to a native Python type. SQL databases often have a few numeric types, including integers and floating-point numbers. Some databases will handle decimal values that map to Python's decimal.Decimal class; this isn't a universal capability, and some databases force the application to convert between decimal.Decimal and text to avoid the truncation problems inherent with floating-point values.

The second important distinction is the tempo of change. A SQL database schema tends to change slowly, and change often involves a review of the impact of the change. In some cases, CSV files are built by interactive spreadsheet software, and manual operations are used to create and save the data. Unsurprisingly, the interactive use of spreadsheets leads to small changes and inconsistencies over short periods of time. While some CSV files are produced by highly automated tools, there may be less scrutiny applied to the order or names of columns.

A third important distinction relates to the design of spreadsheets contrasted with the design of a database. A relational database is often highly normalized; this is an attempt to avoid redundancy. Rather than repeat a group of related values, an entity is assigned to a separate table with a primary key. References to the group of values via the primary key are used to avoid repetition of the values themselves. It's less common to apply normalization rules to a spreadsheet.

Because spreadsheet data may not be fully normalized, extracting meaningful data from a spreadsheet often becomes a rather complicated problem. This can be exacerbated when spreadsheets are tweaked manually or the design of the spreadsheet changes suddenly. To reflect this, the designs in this book suggest using a hierarchy of classes — or collection of related functions — to build a useful Python object from a spreadsheet row. It is often necessary to keep a large pool of builders available to handle variant spreadsheet data as part of historical analysis.

The designs shown earlier had a `PairBuilder` subclass to create individual sample objects. These designs used an `Extract` class to manage the overall construction of samples from the source file. This provided flexibility to handle spreadsheet data.

A database extract is somewhat less likely to need a flexible hierarchy of objects to create useful Python objects. Instead, the needed flexibility is often implemented by changing SQL statements to reflect schema changes or a deeper understanding of the available data. For this reason, we encourage the use of a TOML-format file to keep the SQL statements, permitting some changes without having to add more subclasses to the Python application. The TOML-format configuration files can have version numbers in the file name (and in the comments) to make it clear which database schema they are designed against.

Now that we have a design approach, it's important to make sure we have a list of deliverables that serve as a definition of "Done."

Deliverables

This project has the following deliverables:

- Documentation in the `docs` folder.

- Acceptance tests in the `tests/features` and `tests/steps` folders.

- The acceptance tests will involve creating and destroying example databases as test fixtures.

- Unit tests for application modules in the `tests` folder.

- Mock objects for the database connection will be part of the unit tests.

- Application to acquire data from a SQL database.

We'll look at a few of these deliverables in a little more detail.

Mock database connection and cursor objects for testing

For the data acquisition application, it's essential to provide a mock connection object to expose the SQL and the parameters that are being provided to the database. This mock object can also provide a mock cursor as a query result.

As noted earlier in *Deliverables*, this means the connection object should be created only in the `main()` function. It also means the connection object should be a parameter to any other functions or methods that perform database operations. If the connection object is referenced consistently, it becomes easier to test by providing a mock connection object.

We'll look at this in two parts: first, the conceptual Given and When steps; after that, we'll look at the Then steps. This is sometimes called "arrange-act-assert". Here's the start of the **PyTest** test case:

```python
import sqlite3
from typing import Any, cast
from unittest.mock import Mock, call, sentinel
from pytest import fixture
import db_extract
import model


def test_build_sample(
        mock_connection: sqlite3.Connection,
        mock_config: dict[str, Any]
```

```
):
    extract = db_extract.Extract()
    results = list(
        extract.series_iter(mock_connection, mock_config)
    )
```

The assertions confirm the results come from the mock objects without being transformed, dropped, or corrupted by some error in the code under test. The assertions look like this example:

```
assert results == [
    model.Series(
        name=sentinel.Name,
        samples=[
            model.SeriesSample(sentinel.X, sentinel.Y)
        ]
    )
]
assert cast(Mock, mock_connection).execute.mock_calls == [
    call(sentinel.Names_Query),
    call(sentinel.Samples_Query, {'name': sentinel.Name})
]
```

A mock connection object must provide results with sentinel objects that have the proper structure to look like the iterable `Cursor` object that is returned by SQLite3 when executing a database query.

The mock connection seems rather complicated because it involves two separate mock cursors and a mock connection. Here's some typical code for a mock connection:

```
@fixture
def mock_connection() -> sqlite3.Connection:
    names_cursor: list[tuple[Any, ...]] = [
```

```
        (sentinel.Name,)
    ]
    samples_cursor: list[tuple[Any, ...]]  = [
        (sentinel.X, sentinel.Y)
    ]
    query_to_cursor: dict[sentinel, list[tuple[Any, ...]]] = {
        sentinel.Names_Query: names_cursor,
        sentinel.Samples_Query: samples_cursor
    }


    connection = Mock(
        execute=Mock(
            side_effect=lambda query, param=None: query_to_cursor[query]
        )
    )
    return cast(sqlite3.Connection, connection)
```

The mocked cursors are provided as simple lists. If the code under test used other features of a cursor, a more elaborate Mock object would be required. The query_to_cursor mapping associates a result with a particular query. The idea here is the queries will be sentinel objects, not long SQL strings.

The connection object uses the side-effect feature of Mock objects. When the execute() method is evaluated, the call is recorded, and the result comes from the side-effect function. In this case, it's a lambda object that uses the query_to_cursor mapping to locate an appropriate cursor result.

This use of the side-effect feature avoids making too many assumptions about the internal workings of the unit under test. The SQL will be a sentinel object and the results will contain sentinel objects.

In this case, we're insisting the unit under test does no additional processing on the values retrieved from the database. In other applications, where additional processing is being done, more sophisticated mock objects or test literals may be required.

It's not unusual to use something like (11, 13) instead of (sentinel.X, sentinel.Y) to check that a computation is being performed correctly. However, it's more desirable to isolate the computations performed on SQL results into separate functions. This allows testing these functions as separate units. The SQL retrieval processing can be tested using mock functions for these additional computations.

Also, note the use of the cast() function from the typing module to tell tools like **mypy** this object can be used like a Connection object.

Unit test for a new acquisition module

Throughout this sequence of chapters, the overall acquisition module has grown more flexible. The idea is to permit a wide variety of data sources for an analysis project.

Pragmatically, it is more likely to modify an application to work with a number of distinct CSV formats, or a number of distinct database schemas. When a RESTful API changes, it's often a good strategy to introduce new classes for the changed API as an alternatives to existing classes. Simply modifying or replacing the old definition — in a way — erases useful history on why and how an API is expected to work. This is the Open/Closed principle from the SOLID design principles: the design is open to extension but closed to modification.

Acquiring data from a wide variety of data sources — as shown in these projects — is less likely than variations in a single source. As an enterprise moves from spreadsheets to central databases and APIs, then the analytical tools should follow the data sources.

The need for flexible data acquisition drives the need to write unit tests for the acquisition module to both cover the expected cases and cover the potential domain of errors and mistakes in use.

Acceptance tests using a SQLite database

The acceptance tests need to create (and destroy) a test database. The tests often need to create, retrieve, update, and delete data in the test database to arrange data for the given step or assert the results in the Then step.

In the context of this book, we started with the *Project 1.4: A local SQL database* project to build a test database. There aren't many readily accessible, public, relational databases with extractable data. In most cases, these databases are wrapped with a RESTful API.

The database built in the previous project has two opposing use cases:

- It is for test purposes and can be deleted and rebuilt freely.

- This database must be treated as if it's precious enterprise data, and should not be deleted or updated.

> When we think of the database created in *Project 1.4: A local SQL database* as if it were production data, we need to protect it from unexpected changes.
>
> This means our acceptance tests must build a separate, small, test database, separate from the "production" database created by the previous project.
>
> The test database must not collide with precious enterprise data.

There are two common strategies to avoid collisions between test databases and enterprise databases:

1. Use OS-level security in the file system to make it difficult to damage the files that comprise a shared database. Also, using strict naming conventions can put a test database into a separate namespace that won't collide with production databases.

2. Run the tests in a **Docker container** to create a virtual environment in which production data cannot be touched.

As we noted above in *Approach*, the idea behind a database involves two containers:

- A container for the application components that extract the data.

- A container for the database components that provide the data. An acceptance test can create an ephemeral database service.

With SQLite, however, there is no distinct database service container. The database components become part of the application's components and run in the application's container. The lack of a separate service container means SQLite breaks the conceptual two-container model that applies to large, enterprise databases. We can't create a temporary, mock database **service** for testing purposes.

Because the SQLite database is nothing more than a file, we must focus on OS-level permissions, file-system paths, and naming conventions to keep our test database separate from the production database created in an earlier project. We emphasize this because working with a more complicated database engine (like MySQL or PostgreSQL) will also involve the same consideration of permissions, file paths, and naming conventions. Larger databases will add more considerations, but the foundations will be similar.

 It's imperative to avoid disrupting production operations while creating data analytic applications.

Building and destroying a temporary SQLite database file suggests the use of a `@fixture` to create a database and populate the needed schema of tables, views, indexes, etc. The Given steps of individual scenarios can provide a summary of the data arrangement required by the test.

We'll look at how to define this as a feature. Then, we can look at the steps required for the implementation of the fixture, and the step definitions.

The feature file

Here's the kind of scenario that seems to capture the essence of a SQL extract application:

```
@fixture.sqlite
Scenario: Extract data from the enterprise database
```

```
Given a series named "test1"

And sample values "[(11, 13), (17, 19)]"

When we run the database extract command with the test fixture database

Then log has INFO line with "series: test1"

And log has INFO line with "count: 2"

And output directory has file named "quartet/test1.csv"
```

The `@fixture`. tag follows the common naming convention for associating specific, reusable fixtures with scenarios. There are many other purposes for tagging scenarios in addition to specifying the fixture to use. In this case, the fixture information is used to build an SQLite database with an empty schema.

The Given steps provide some data to load into the database. For this acceptance test, a single series with only a few samples is used.

The tag information can be used by the **behave** tool. We'll look at how to write a `before_tag()` function to create (and destroy) the temporary database for any scenario that needs it.

The sqlite fixture

The fixture is generally defined in the environment.py module that the **behave** tool uses. The `before_tag()` function is used to process the tags for a feature or a scenario within a feature. This function lets us then associate a specific feature function with the scenario:

```python
from behave import fixture, use_fixture
from behave.runner import Context


def before_tag(context: Context, tag: str) -> None:
    if tag == "fixture.sqlite":
        use_fixture(sqlite_database, context)
```

The `use_fixture()` function tells the **behave** runner to invoke the given function, `sqlite_database()`, with a given argument value – in this case, the context object. The `sqlite_database()` function should be a generator: it can prepare the database, execute a `yield` statement, and then destroy the database. The **behave** runner will consume the yielded value as part of setting up the test, and the consume one more value when it's time to tear down the test.

The function to create (and destroy) the database has the following outline:

```python
from collections.abc import Iterator
from pathlib import Path
import shutil
import sqlite3
from tempfile import mkdtemp
import tomllib


from behave import fixture, use_fixture
from behave.runner import Context


@fixture
def sqlite_database(context: Context) -> Iterator[str]:
    # Setup: Build the database files (shown later).


    yield context.db_uri


    # Teardown: Delete the database files (shown later).
```

We've decomposed this function into three parts: the setup, the `yield` to allow the test scenario to proceed, and the teardown. We'll look at the *Set up: Build the database files* and the *Teardown: Delete the database files* sections separately.

The setup processing of the `sqlite_database()` function is shown in the following snippet:

```python
    # Get Config with SQL to build schema.
    config_path = Path.cwd() / "schema.toml"
```

```python
with config_path.open() as config_file:

    config = tomllib.load(config_file)

    create_sql = config['definition']['create']

    context.manipulation_sql = config['manipulation']

# Build database file.

context.working_path = Path(mkdtemp())

context.db_path = context.working_path / "test_example.db"

context.db_uri = f"file:{context.db_path}"

context.connection = sqlite3.connect(context.db_uri, uri=True)

for stmt in create_sql:

    context.connection.execute(stmt)

context.connection.commit()
```

The configuration file is read from the current working directory. The SQL statements to create the database and perform data manipulations are extracted from the schema. The database creation SQL will be executed during the tag discovery. The manipulation SQL will be put into the context for use by the Given steps executed later.

Additionally, the context is loaded up with a working path, which will be used for the database file as well as the output files. The context will have a db_uri string, which can be used by the data extract application to locate the test database.

Once the context has been filled, the individual SQL statements can be executed to build the empty database.

After the yield statement, the teardown processing of the sqlite_database() function is shown in the following snippet:

```python
context.connection.close()
shutil.rmtree(context.working_path)
```

The SQLite3 database must be closed before the files can be removed. The shutil package includes functions that work at a higher level on files and directories. The rmtree() function removes the entire directory tree and all of the files within the tree.

This fixture creates a working database. We can now write step definitions that depend on this fixture.

The step definitions

We'll show two-step definitions to insert series and samples into the database. The following example shows the implementation of one of the Given steps:

```
@given(u'a series named "{name}"')
def step_impl(context, name):
    insert_series = context.manipulation_sql['insert_series']
    cursor = context.connection.execute(
        insert_series,
        {'series_id': 99, 'name': name}
    )
    context.connection.commit()
```

The step definition shown above uses SQL to create a new row in the `series` table. It uses the connection from the context; this was created by the `sqlite_database()` function that was made part of the testing sequence by the `before_tag()` function.

The following example shows the implementation of the other Given step:

```
@given(u'sample values "{list_of_pairs}"')
def step_impl(context, list_of_pairs):
    pairs = literal_eval(list_of_pairs)
    insert_values = context.manipulation_sql['insert_values']
    for seq, row in enumerate(pairs):
        cursor = context.connection.execute(
            insert_values,
            {'series_id': 99, 'sequence': seq, 'x': row[0], 'y': row[1]}
        )
    context.connection.commit()
```

The step definition shown above uses SQL to create a new row in the `series_sample` table. It uses the connection from the context, also.

Once the series and samples have been inserted into the database, the `When` step can run the data acquisition application using the database URI information from the context.

The `Then` steps can confirm the results from running the application match the database seeded by the fixture and the `Given` steps.

With this testing framework in place, you can run the acceptance test suite. It's common to run the acceptance tests before making any of the programming changes; this reveals the `acquire` application doesn't pass all of the tests.

In the next section, we'll look at the database extract module and rewrite the main application.

The Database extract module, and refactoring

This project suggests three kinds of changes to the code written for the previous projects:

- Revise the `model` module to expand on what a "series" is: it's a parent object with a name and a list of subsidiary objects.

- Add the `db_extract` module to grab data from a SQL database.

- Update the `acquire` module to gather data from any of the available sources and create CSV files.

Refactoring the `model` module has a ripple effect on other projects, requiring changes to those modules to alter the data structure names.

As we noted in *Approach*, it's common to start a project with an understanding that evolves and grows. More exposure to the problem domain, the users, and the technology shifts our understanding. This project reflects a shift in understanding and leads to a need to change the implementation of previously completed projects.

One consequence of this is exposing the series' name. In projects from previous chapters, the four series had names that were arbitrarily imposed by the application program. Perhaps a file name might have been `"series_1.csv"` or something similar.

Working with the SQL data exposed a new attribute, the name of a series. This leads to two profound choices for dealing with this new attribute:

1. Ignore the new attribute.

2. Alter the previous projects to introduce a series name.

Should the series name be the file name? This seems to be a bad idea because the series name may have spaces or other awkward punctuation.

It seems as though some additional metadata is required to preserve the series name and associate series names with file names. This would be an extra file, perhaps in JSON or TOML format, created as part of the extract operation.

Summary

This chapter's projects covered two following essential skills:

- Building SQL databases. This includes building a representative of a production database, as well as building a test database.

- Extracting data from SQL databases.

This requires learning some SQL, of course. SQL is sometimes called the *lingua franca* of data processing. Many organizations have SQL databases, and the data must be extracted for analysis.

Also important is learning to work in the presence of precious production data. It's important to consider the naming conventions, file system paths, and permissions associated with database servers and the files in use. Attempting to extract analytic data is not a good reason for colliding with production operations.

The effort required to write an acceptance test that uses an ephemeral database is an important additional skill. Being able to create databases for test purposes permits debugging by identifying problematic data, creating a test case around it, and then working in an isolated development environment. Further, having ephemeral databases permits examining changes to a production database that might facilitate analysis or resolve uncertainty in production data.

In the next chapter, we'll transition from the bulk acquisition of data to understanding the relative completeness and usefulness of the data. We'll build some tools to inspect the raw data that we've acquired.

Extras

Here are some ideas for the reader to add to this project.

Consider using another database

For example, MySQL or PostgreSQL are good choices. These can be downloaded and installed on a personal computer for non-commercial purposes. The administrative overheads are not overly burdensome.

It is essential to recognize these are rather large, complex tools. For readers new to SQL, there is a lot to learn when trying to install, configure, and use one of these databases.

See `https://dev.mysql.com/doc/mysql-getting-started/en/` for some advice on installing and using MySQL.

See `https://www.postgresql.org/docs/current/tutorial-start.html` for advice on installing and using PostgreSQL.

In some cases, it makes sense to explore using a Docker container to run a database server on a virtual machine. See `https://www.packtpub.com/product/docker-for-develop ers/9781789536058` for more about using Docker as a way to run complex services in isolated environments.

See `https://dev.mysql.com/doc/refman/8.0/en/docker-mysql-getting-started.htm l` for ways to use MySQL in a Docker container.

See `https://www.docker.com/blog/how-to-use-the-postgres-docker-official-ima`

ge/ for information on running PostgreSQL in a Docker container.

Consider using a NoSQL database

A NoSQL database offers many database features — including reliably persistent data and shared access — but avoids (or extends) the relational data model and replaces the SQL language.

This leads to data acquisition applications that are somewhat like the examples in this chapter. There's a connection to a server and requests to extract data from the server. The requests aren't SQL SELECT statements. Nor is the result necessarily rows of data in a completely normalized structure.

For example, MongoDB. Instead of rows and tables, the data structure is JSON documents. See https://www.packtpub.com/product/mastering-mongodb-4x-second-edition/9 781789617870.

The use of MongoDB changes data acquisition to a matter of locating the JSON documents and then building the desired document from the source data in the database.

This would lead to two projects, similar to the two described in this chapter, to populate the "production" Mongo database with some data to extract, and then writing the acquisition program to extract the data from the database.

Another alternative is to use the PostgreSQL database with JSON objects for the data column values. This provides a MongoDB-like capability using the PostgreSQL engine. See https://www.postgresql.org/docs/9.3/functions-json.html for more information on this approach.

Here are some common categories of NoSQL databases:

- Document databases

- Key-value stores

- Column-oriented databases

- Graph databases

The reader is encouraged to search for representative products in these categories and consider the two parts of this chapter: loading a database and acquiring data from the database.

Consider using SQLAlchemy to define an ORM layer

In *The Object-Relational Mapping (ORM) problem* we talked about the ORM problem. In that section, we made the case that using a tool to configure an ORM package for an existing database can sometimes turn out badly.

This database, however, is very small. It's an ideal candidate for learning about simple ORM configuration.

We suggest starting with the SQLAlchemy ORM layer. See `https://docs.sqlalchemy.org/en/20/orm/quickstart.html` for advice on configuring class definitions that can be mapped to tables. This will eliminate the need to write SQL when doing extracts from the database.

There are other ORM packages available for Python, also. The reader should feel free to locate an ORM package and build the extraction project in this chapter using the ORM data model.

6

Project 2.1: Data Inspection Notebook

We often need to do an ad hoc inspection of source data. In particular, the very first time we acquire new data, we need to see the file to be sure it meets expectations. Additionally, debugging and problem-solving also benefit from ad hoc data inspections. This chapter will guide you through using a Jupyter notebook to survey data and find the structure and domains of the attributes.

The previous chapters have focused on a simple dataset where the data types look like obvious floating-point values. For such a trivial dataset, the inspection isn't going to be very complicated.

It can help to start with a trivial dataset and focus on the tools and how they work together. For this reason, we'll continue using relatively small datasets to let you learn about the tools without having the burden of **also** trying to understand the data.

This chapter's projects cover how to create and use a Jupyter notebook for data inspection.

This permits tremendous flexibility, something often required when looking at new data for the first time. It's also essential when diagnosing problems with data that has — unexpectedly — changed.

A Jupyter notebook is inherently interactive and saves us from having to design and build an interactive application. Instead, we need to be disciplined in using a notebook only to examine data, never to apply changes.

This chapter has one project, to build an inspection notebook. We'll start with a description of the notebook's purpose.

Description

When confronted with raw data acquired from a source application, database, or web API, it's prudent to inspect the data to be sure it really can be used for the desired analysis. It's common to find that data doesn't precisely match the given descriptions. It's also possible to discover that the metadata is out of date or incomplete.

The foundational principle behind this project is the following:

We don't always know what the actual data looks like.

Data may have errors because source applications have bugs. There could be "undocumented features," which are similar to bugs but have better explanations. There may have been actions made by users that have introduced new codes or status flags. For example, an application may have a "comments" field on an accounts-payable record, and accounting clerks may have invented their own set of coded values, which they put in the last few characters of this field. This defines a manual process outside the enterprise software. It's an essential business process that contains valuable data; it's not part of any software.

The general process for building a useful Jupyter notebook often proceeds through the following phases:

1. Start with a simple display of selected rows.

2. Then, show ranges for what appear to be numerical fields.

3. Later, in a separate analysis notebook, we can find central tendency (mean, median, and standard deviation) values after they've been cleaned.

Using a notebook moves us away from the previous chapters' focus on CLI applications. This is necessary because a notebook is interactive. It is designed to allow exploration with few constraints.

The **User Experience (UX)** has two general steps to it:

1. Run a data acquisition application. This is one of the CLI commands for projects in any of the previous chapters.

2. Start a Jupyter Lab server. This is a second CLI command to start the server. The `jupyter lab` command will launch a browser session. The rest of the work is done through the browser:

   (a) Create a notebook by clicking the notebook icon.

   (b) Load data by entering some Python code into a cell.

   (c) Determine if the data is useful by creating cells to show the data and show properties of the data.

For more information on Jupyter, see `https://www.packtpub.com/product/learning-jupyter/9781785884870`.

About the source data

An essential ingredient here is that all of the data acquisition projects **must** produce output in a consistent format. We've suggested using NDJSON (sometimes called JSON NL) as a format for preserving the raw data. See *Chapter 3, Project 1.1: Data Acquisition Base Application*, for more information on the file format.

It's imperative to review the previous projects' acceptance test suites to be sure there is a test to confirm the output files have the correct, consistent format.

To recap the data flow, we've done the following:

- Read from some source. This includes files, RESTful APIs, HTML pages, and SQL databases.

- Preserved the raw data in an essentially textual form, stripping away any data type information that may have been imposed by a SQL database or RESTful JSON document.

The inspection step will look at the text versions of values in these files. Later projects, starting with *Chapter 9, Project 3.1: Data Cleaning Base Application*, will look at converting data from text into something more useful for analytic work.

An inspection notebook will often be required to do some data cleanup in order to show data problems. This will be enough cleanup to understand the data and no more. Later projects will expand the cleanup to cover all of the data problems.

In many data acquisition projects, it's unwise to attempt any data conversion before an initial inspection. This is because the data is highly variable and poorly documented. A disciplined, three-step approach separates acquisition and inspection from attempts at data conversion and processing.

We may find a wide variety of unexpected things in a data source. For example, a CSV file may have an unexpected header, leading to a row of bad data. Or, a CSV file may — sometimes — lack headers, forcing the acquire application to supply default headers. A file that's described as CSV may not have delimiters, but may have fixed-size text fields padded with spaces. There may be empty rows that can be ignored. There may be empty rows that delimit the useful data and separate it from footnotes or other non-data in the file. A ZIP archive may contain a surprising collection of irrelevant files in addition to the desired data file.

Perhaps one of the worst problems is trying to process files that are not prepared using a widely used character encoding such as UTF-8. Files encoded with CP-1252 encoding may have a few odd-looking characters when the decoder assumes it's UTF-8 encoding.

Python's `codecs` module provides a number of forms of alternative file encoding to handle this kind of problem. This problem seems rare; some organizations will note the encoding for text to prevent problems.

Inspection notebooks often start as `print()` functions in the data acquisition process to show what the data is. The idea here is to extend this concept a little and use an interactive notebook instead of `print()` to get a look at the data and see that it meets expectations.

> Not all managers agree with taking time to build an inspection notebook. Often, this is a conflict between assumptions and reality with the following potential outcomes:
>
> - A manager can assume there will be no surprises in the data; the data will be entirely as specified in a data contract or other schema definition.
>
>   - When the data doesn't match expectations, a data inspection notebook will be a helpful part of the debugging effort.
>
>
>
>   - In the unlikely event the data does match expectations, the data inspection notebook can be used to show that the data is valid.
>
> - A manager can assume the data is unlikely to be correct. In this case, the data inspection notebook will be seen as useful for uncovering the inevitable problems.
>
> Notebooks often start as `print()` or logger output to confirm the data is useful. This debugging output can be migrated to an informal notebook — at a low cost — and evolve into something more complete and focused on inspection and data quality assurance.

This initial project won't build a complicated notebook. The intent is to provide an **interactive** display of data, allowing exploration and investigation. In the next section, we'll outline an approach to this project, and to working with notebooks in general.

Approach

We'll take some guidance from the C4 model (`https://c4model.com`) when looking at our approach.

- **Context**: For this project, the context diagram has two use cases: acquire and inspect

- **Containers**: There's one container for the various applications: the user's personal computer

- **Components**: There are two significantly different collections of software components: the acquisition program and inspection notebooks

- **Code**: We'll touch on this to provide some suggested directions

A context diagram for this application is shown in *Figure 6.1*.

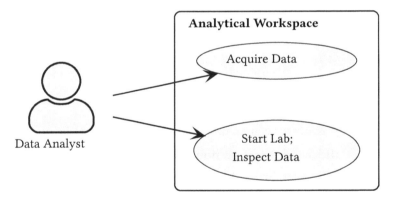

Figure 6.1: Context Diagram

The data analyst will use the CLI to run the data acquisition program. Then, the analyst will use the CLI to start a Jupyter Lab server. Using a browser, the analyst can then use Jupyter Lab to inspect the data.

The components fall into two overall categories. The component diagram is shown in *Figure 6.2*.

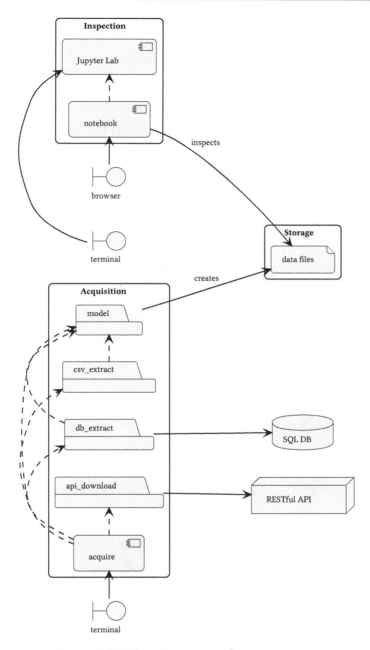

Figure 6.2: Component diagram

The diagram shows the interfaces seen by the data analyst, the `terminal` and the `browser`.

These are shown with the *boundary* icon from the **Unified Modeling Language (UML)**.

The `Acquisition` group of components contains the various modules and the overall acquire application. This is run from the command line to acquire the raw data from an appropriate source. The `db_extract` module is associated with an external SQL database. The `api_download` module is associated with an external RESTful API. Additional sources and processing modules could be added to this part of the diagram.

The processing performed by the `Acquisition` group of components creates the data files shown in the `Storage` group. This group depicts the raw data files acquired by the `acquire` application. These files will be refined and processed by further analytic applications.

The `Inspection` group shows the `jupyter` component. This is the entire Jupyter Lab application, summarized as a single icon. The `notebook` component is the notebook we'll build in this application. This notebook depends on Jupyter Lab.

The `browser` is shown with the boundary icon. The intention is to characterize the notebook interaction via the browser as the user experience.

The `notebook` component will use a number of built-in Python modules. This notebook's cells can be decomposed into two smaller kinds of components:

- Functions to gather data from acquisition files.

- Functions to show the raw data. The `collections.Counter` class is very handy for this.

You will need to locate (and install) a version of Jupyter Lab for this project. This needs to be added to the `requirements-dev.txt` file so other developers know to install it.

When using `conda` to manage virtual environments, the command might look like the following:

```
% conda install jupyterlab
```

When using other tools to manage virtual environments, the command might look like the following:

```
% python -m pip install jupyterlab
```

Once the `jupyter` products are installed, it must be started from the command line. This command will start the server and launch a browser window:

```
% jupyter lab
```

For information on using Jupyter Lab, see `https://jupyterlab.readthedocs.io/en/latest/`.

If you're not familiar with Jupyter, now is the time to use tutorials and learn the basics before moving on with this project.

Many of the notebook examples will include `import` statements.

Developers new to working with a Jupyter notebook should not take this as advice to repeat `import` statements in multiple cells throughout a notebook.

In a practical notebook, the imports can be collected together, often in a separate cell to introduce all of the needed packages.

In some enterprises, a startup script is used to provide a common set of imports for a number of closely related notebooks.

We'll return to more flexible ways to handle Python libraries from notebooks in *Chapter 13, Project 4.1: Visual Analysis Techniques.*

There are two other important considerations for this project: the ability to write automated tests for a notebook and the interaction of Python modules and notebooks. We'll look at these topics in separate sections.

Notebook test cases for the functions

It's common to require unit test cases for a Python package. To be sure the test cases are meaningful, some enterprises insist the test cases exercise 100% of the lines of code in the module. For some industries, all logic paths must be tested. For more information, see *Chapter 1, Project Zero: A Template for Other Projects*.

For notebooks, automated testing can be a shade more complicated than it is for a Python module or package. The complication is that notebooks can contain arbitrary code that is not designed with testability in mind.

In order to have a disciplined, repeatable approach to creating notebooks, it's helpful to develop a notebook in a series of stages, evolving toward a notebook that supports automated testing.

A notebook is software, and without test cases, any software is untrustworthy. In rare cases, the notebook's code is simple enough that we can inspect it to develop some sense of its overall fitness for purpose. In most cases, complex computations, functions, and class definitions require a test case to demonstrate the code can be trusted to work properly.

The stages of notebook evolution often work as follows:

0 At stage zero, a notebook is often started with arbitrary Python code in cells and few or no function or class definitions. This is a great way to start development because the interactive nature of the notebook provides immediate results. Some cells will have errors or bad ideas in them. The order for processing the cells is not simply top to bottom. This code is difficult (or impossible) to validate with any automated testing.

1 Stage one will transform cell expressions into function and class definitions. This version of the notebook can also have cells with examples using the functions and classes. The order is closer to strictly top-to-bottom; there are fewer cells with known errors. The presence of examples serves as a basis for validating the notebook's processing, but an automated test isn't available.

2 Stage two has more robust tests, using formal `assert` statements or `doctest` comments to define a repeatable test procedure. Rerunning the notebook from the beginning after any change will validate the notebook by executing the `assert` statements. All the cells are valid and the notebook processing is strictly top to bottom.

3 When there is more complicated or reusable processing, it may be helpful to refactor the function and class definitions out of the notebook and into a module. The module will have a unit test module or may be tested via **doctest** examples. This new module will be imported by the notebook; the notebook is used more for the presentation of results than the development of new ideas.

One easy road to automated testing is to include **doctest** examples inside function and class definitions. For example, we might have a notebook cell that contains something like the following function definition:

```python
def min_x(series: Series) -> float:
    """
    >>> s = [
    ...     {'x': '3', 'y': '4'},
    ...     {'x': '2', 'y': '3'},
    ...     {'x': '5', 'y': '6'}]
    >>> min_x(s)
    2
    """
    return min(int(s['x']) for s in series.samples)
```

The lines in the function's docstring marked with >>> are spotted by the **doctest** tool. The lines are evaluated and the results are compared with the example from the docstring.

The last cell in the notebook can execute the `doctest.testmod()` function. This will examine all of the class and function definitions in the notebook, locate their **doctest** examples, and confirm the actual results match the expectations.

For additional tools to help with notebook testing, see `https://testbook.readthedocs.i`

```
o/en/latest/.
```

This evolution from a place for recording good ideas to an engineered solution is not trivially linear. There are often exploration and learning opportunities that lead to changes and shifts in focus. Using a notebook as a tool for tracking ideas, both good and bad, is common.

A notebook is also a tool for presenting a final, clear picture of whether or not the data is what the users expect. In this second use case, separating function and class definitions becomes more important. We'll look at this briefly in the next section.

Common code in a separate module

As we noted earlier, a notebook lets an idea evolve through several forms.

We might have a cell with the following

```
x_values = []
for s in source_data[1:]:
    x_values.append(float(s['x']))
min(x_values)
```

Note that this computation skips the first value in the series. This is because the source data has a header line that's read by the csv.reader() function. Switching to csv.DictReader() can politely skip this line, but also changes the result structure from a list of strings into a dictionary.

This computation of the minimum value can be restated as a function definition. Since it does three things — drops the first line, extracts the 'x' attribute, and converts it into a float — it might be better to decompose it into three functions. It can be refactored again to include **doctest** examples in each function. See *Notebook test cases for the functions* for the example.

Later, this function can be cut from the notebook cell and pasted into a separate module. We'll assume the overall function was named min_x(). We might add this to a module named `series_stats.py`. The notebook can then import and use the function, leaving the definition as a sidebar detail:

```
from series_stats import min_x
```

When refactoring a notebook to a reusable module, it's important to use **cut and paste**, not copy and paste. A copy of the function will lead to questions if one of the copies is changed to improve performance or fix a problem and the other copy is left untouched. This is sometimes called the **Don't Repeat Yourself (DRY)** principle.

When working with external modules that are still under development, any changes to a module will require stopping the notebook kernel and rerunning the notebook from the very beginning to remove and reload the function definitions. This can become awkward. There are some extensions to iPython that can be used to reload modules, or even auto-reload modules when the source module changes.

An alternative is to wait until a function or class seems mature and unlikely to change before refactoring the notebook to create a separate module. Often, this decision is made as part of creating a final presentation notebook to display useful results.

We can now look at the specific list of deliverables for this project.

Deliverables

This project has the following deliverables:

- A `pyproject.toml` file that identifies the tools used. For this book, we used `jupyterlab==3.5.3`. Note that while the book was being prepared for publication, version 4.0 was released. This ongoing evolution of components makes it important for you to find the latest version, not the version quoted here.

- Documentation in the `docs` folder.

- Unit tests for any new application modules in the `tests` folder.

- Any new application modules in the `src` folder with code to be used by the inspection notebook.

- A notebook to inspect the raw data acquired from any of the sources.

The project directory structure suggested in *Chapter 1, Project Zero: A Template for Other Projects* mentions a `notebooks` directory. See *List of deliverables* for more information. Previous chapters haven't used any notebooks, so this directory might not have been created in the first place. For this project, the `snotebooks` directory is needed.

Let's look at a few of these deliverables in a little more detail.

Notebook .ipynb file

The notebook can (and should) be a mixture of Markdown cells providing notes and context, and computation cells showing the data.

Readers who have followed the projects up to this point will likely have a directory with NDJSON files that need to be read to construct useful Python objects. One good approach is to define a function to read lines from a file, and use `json.loads()` to transform the line of text into a small dictionary of useful data.

There's no compelling reason to use the `model` module's class definitions for this inspection. The class definitions can help to make the data somewhat more accessible.

The inspection process starts with cells that name the files, creating `Path` objects.

A function code like the following example might be helpful:

```python
import csv
from collections.abc import Iterator
import json
from typing import TextIO


def samples_iter(source: TextIO) -> Iterator[dict[str, str]]:
    yield from (json.loads(line) for line in source)
```

This function will iterate over the acquired data. In many cases, we can use the iterator to scan through a large collection of samples, picking individual attribute values or some subset of the samples.

We can use the following statement to create a list-of-dictionary structure from the given path:

```
from pathlib import Path
source_path = Path("/path/to/quartet/Series_1.ndjson")
with source_path.open() as source_file:
    source_data = list(samples_iter(source_file))
```

We can start with these basics in a few cells of the notebook. Given this foundation, further cells can explore the available data.

Cells and functions to analyze data

For this initial inspection project, the analysis requirements are small. The example datasets from the previous chapters are artificial data, designed to demonstrate the need to use graphical techniques for exploratory data analysis.

For other datasets, however, there may be a variety of odd or unusual problems.

For example, the **CO$_2$ PPM — Trends in Atmospheric Carbon Dioxide** dataset, available at `https://datahub.io/core/co2-ppm`, has a number of "missing value" codes in the data. Here are two examples:

- The CO$_2$ average values sometimes have values of −99.99 as a placeholder for a time when a measurement wasn't available. In these cases, a statistical process used data from adjacent months to interpolate the missing value.

- Additionally, the number of days of valid data for a month's summary wasn't recorded, and a −1 value is used.

This dataset requires a bit more care to be sure of the values in each column and what the columns mean.

Capturing the domain of values in a given column is helpful here. The `collections` module has a `Counter` object that's ideal for understanding the data in a specific column.

A cell can use a three-step computation to see the domain of values:

1. Use the `samples_iter()` function to yield the source documents.

2. Create a generator with sample attribute values.

3. Create a `Counter` to summarize the values.

This can lead to a cell in the notebook with the following statements:

```
from collections import Counter


values_x = (sample['x'] for sample in source_data)
domain_x = Counter(values_x)
```

The next cell in the notebook can display the value of the `domain_x` value. If the `csv.reader()` function is used, it will reveal the header along with the domain of values. If the `csv.DictReader()` class is used, this collection will not include the header. This permits a tidy exploration of the various attributes in the collection of samples.

An inspection notebook is not the place to attempt more sophisticated data analysis. Computing means or medians should only be done on cleaned data. We'll return to this in *Chapter 15, Project 5.1: Modeling Base Application.*

Cells with Markdown to explain things

It's very helpful to include cells using Markdown to provide information, insights, and lessons learned about the data.

For information on the markdown language, see the Daring Fireball website: `https://daringfireball.net/projects/markdown/basics`.

As noted earlier in this chapter, there are two general flavors of notebooks:

- **Exploratory**: These notebooks are a series of blog posts about the data and the process of exploring and inspecting the data. Cells may not all work because they're works in process.

- **Presentation**: These notebooks are a more polished, final report on data or problems. The paths that lead to dead ends should be pruned into summaries of the lessons learned.

A bright line separates these two flavors of notebooks. The distinguishing factor is the reproducibility of the notebook. A notebook that's useful for presentations can be run from beginning to end without manual intervention to fix problems or skip over cells with syntax errors or other problems. Otherwise, the notebook is part of an exploration. It's often necessary to copy and edit an exploratory notebook to create a derived notebook focused on presentation.

Generally, a notebook designed for a presentation uses Markdown cells to create a narrative flow that looks like any chapter of a book or article in a journal. We'll return to more formal reporting in *Chapter 14, Project 4.2: Creating Reports*.

Cells with test cases

Earlier, we introduced a `samples_iter()` function that lacked any unit tests or examples. It's considerably more helpful to provide a **doctest** string within a notebook:

```python
def samples_iter(source: TextIO) -> Iterator[dict[str, str]]:
    """

    # Build NDJSON file with two lines
    >>> import json
    >>> from io import StringIO
    >>> source_data = [
    ...     {'x': 0, 'y': 42},
    ...     {'x': 1, 'y': 99},
    ... ]
    >>> source_text = [json.dumps(sample) for sample in source_data]
    >>> ndjson_file = StringIO('\\n'.join(source_text))

    # Parse the file
```

```
>>> list(samples_iter(ndjson_file))
[{'x': 0, 'y': 42}, {'x': 1, 'y': 99}]
"""
yield from (json.loads(line) for line in source)
```

This function's docstrings include an extensive test case. The test case builds an NDJSON document from a list of two dictionaries. The test case then applies the `samples_iter()` function to parse the NDJSON file and recover the original two samples.

To execute this test, the notebook needs a cell to examine the docstrings in all of the functions and classes defined in the notebook:

```
import doctest
doctest.testmod()
```

This works because the global context for a notebook is treated like a module with a default name of __main__. This module will be examined by the `textmod()` function to find docstrings that look like they contain doc test examples.

Having the last cell run the **doctest** tool makes it easy to run the notebook, scroll to the end, and confirm the tests have all passed. This is an excellent form of validation.

Executing a notebook's test suite

A Jupyter notebook is inherently interactive. This makes an automated acceptance test of a notebook potentially challenging.

Fortunately, there's a command that executes a notebook to confirm it works all the way through without problems.

We can use the following command to execute a notebook to confirm that all the cells will execute without any errors:

```
% jupyter execute notebooks/example_2.ipynb
```

A notebook may ingest a great deal of data, making it very time-consuming to test the notebook as a whole. This can lead to using a cell to read a configuration file and using this information to use a subset of data for test purposes.

Summary

This chapter's project covered the basics of creating and using a Jupyter Lab notebook for data inspection. This permits tremendous flexibility, something often required when looking at new data for the first time.

We also looked at adding **doctest** examples to functions and running the **doctest** tool in the last cell of a notebook. This lets us validate that the code in the notebook is very likely to work properly.

Now that we've got an initial inspection notebook, we can start to consider the specific kinds of data being acquired. In the next chapter, we'll add features to this notebook.

Extras

Here are some ideas for you to add to this project.

Use pandas to examine data

A common tool for interactive data exploration is the `pandas` package.

See `https://pandas.pydata.org` for more information.

Also, see `https://www.packtpub.com/product/learning-pandas/9781783985128` for resources for learning more about pandas.

The value of using pandas for examining text may be limited. The real value of pandas is for doing more sophisticated statistical and graphical analysis of the data.

We encourage you to load NDJSON documents using pandas and do some preliminary investigation of the data values.

7

Data Inspection Features

There are three broad kinds of data domains: cardinal, ordinal, and nominal. The first project in this chapter will guide you through the inspection of cardinal data; values like weights, measures, and durations where the data is continuous, as well as counts where the data is discrete. The second project will guide reasoners through the inspection of ordinal data involving things like dates, where order matters, but the data isn't a proper measurement; it's more of a code or designator. The nominal data is a code that happens to use digits but doesn't represent numeric values. The third project will cover the more complex case of matching keys between separate data sources.

An inspection notebook is required when looking at new data. It's a great place to keep notes and lessons learned. It's helpful when diagnosing problems that arise in a more mature analysis pipeline.

This chapter will cover a number of skills related to data inspection techniques:

- Essential notebook data inspection features using Python expressions, extended from the previous chapter.

- The `statistics` module for examining cardinal data.

- The `collections.Counter` class for examining ordinal and nominal data.

- Some additional `collections.Counter` for matching primary and foreign keys.

For the Ancombe's Quartet example data set used in *Chapters 3, 4*, and *5*, both of the attribute values are cardinal data. It's a helpful data set for some of the inspections, but we'll need to look at some other data sets for later projects in this chapter. We'll start by looking at some inspection techniques for cardinal data. Readers who are focused on other data sets will need to discern which attributes represent cardinal data.

Project 2.2: Validating cardinal domains — measures, counts, and durations

A great deal of data is cardinal in nature. Cardinal numbers are used to count things, like elements of a set. The concept can be generalized to include real numbers representing a weight or a measure.

A very interesting data set is available here: `https://www.kaggle.com/datasets/rtatman/iris-dataset-json-version`. This contains samples with numerous measurements of the pistils and stamen of different species of flowers. The measurements are identifiable because the unit, mm, is provided.

Another interesting data set is available here: `https://datahub.io/core/co2-ppm`. This contains data with measurements of CO_2 levels measured with units of ppm, parts per million.

We need to distinguish counts and measures from numbers that are only used to rank or order things, which are called ordinal numbers. Also, number-like data is sometimes only a code. US postal codes, for example, are merely strings of digits; they aren't proper numeric values. We'll look at these numeric values in *Project 2.3: Validating text and codes — nominal data and ordinal numbers.*

Since this is an inspection notebook, the primary purpose is only to understand the range of values for cardinal data. A deeper analysis will come later. For now, we want a notebook that demonstrates the data is complete and consistent, and can be used for further processing.

In the event an enterprise is using data contracts, this notebook will demonstrate compliance with the data contract. With data contracts, the focus may shift slightly from showing "some data that is not usable" to showing "data found to be non-compliant with the contract." In cases where the contract is inadequate for the analytical consumer, the notebook may shift further to show "compliant data that's not useful."

We'll start with a description of the kinds of cells to add to an inspection notebook. After that, we'll about the architectural approach and wrap up with a detailed list of deliverables.

Description

This project's intent is to inspect raw data to understand if it is actually cardinal data. In some cases, floating-point values may have been used to represent nominal data; the data appears to be a measurement but is actually a code.

> Spreadsheet software tends to transform all data into floating-point numbers; many data items may look like cardinal data.
>
> One example is US Postal Codes, which are strings of digits, but may be transformed into numeric values by a spreadsheet.
>
> Another example is bank account numbers, which — while very long — can be converted into floating-point numbers. A floating-point value uses 8 bytes of storage, but will comfortably represent about 15 decimal digits. While this is a net saving in storage, it is a potential confusion of data types and there is a (small) possibility of having an account number altered by floating-point truncation rules.

The user experience is a Jupyter Lab notebook that can be used to examine the data, show some essential features of the raw data values, and confirm that the data really does appear to be cardinal.

There are several common sub-varieties of cardinal data:

- Counts; represented by integer values.

- Currency and other money-related values. These are often decimal values, and the `float` type is likely to be a bad idea.

- Duration values. These are often measured in days, hours, and minutes, but represent a time interval or a "delta" applied to a point in time. These can be normalized to seconds or days and represented by a float value.

- More general measures are not in any of the previous categories. These are often represented by floating-point values.

What's important for this project is to have an overview of the data. Later projects will look at cleaning and converting the data for further use. This notebook is only designed to preview and inspect the data.

We'll look at general measures first since the principles apply to counts and durations. Currency, as well as duration, values are a bit more complicated and we'll look at them separately. Date-time stamps are something we'll look at in the next project since they're often thought of as ordinal data, not cardinal.

Approach

This project is based on the initial inspection notebook from *Chapter 6, Project 2.1: Data Inspection Notebook*. Some of the essential cell content will be reused in this notebook. We'll add components to the components shown in the earlier chapter – specifically, the `samples_iter()` function to iterate over samples in an open file. This feature will be central to working with the raw data.

In the previous chapter, we suggested avoiding conversion functions. When starting down the path of inspecting data, it's best to assume nothing and look at the text values first.

There are some common patterns in the source data values:

- The values appear to be all numeric values. The `int()` or `float()` function works on all of the values. There are two sub-cases here:

  - All of the values seem to be proper counts or measures in some expected range. This is ideal.

  - A few "outlier" values are present. These are values that seem to be outside the expected range of values.

- Some of the values are not valid numbers. They may be empty strings, or a code line "NULL", "None", or "N/A".

Numeric outlier values can be measurement errors or an interesting phenomenon buried in the data. Outlier values can also be numeric code values indicating a known missing or otherwise unusable value for a sample. In the example of the CO_2 data, there are outlier values of −99.99 parts per million, which encode a specific kind of missing data situation.

Many data sets will be accompanied by metadata to explain the domain of values, including non-numeric values, as well as the numeric codes in use. Some enterprise data sources will not have complete or carefully explained metadata. This means an analyst needs to ask questions to locate the root cause for non-numeric values or special codes that appear in cardinal data.

The first question — *are all the values numeric?* — can be handled with code like the following:

```python
from collections import defaultdict
from collections.abc import Iterable, Callable
from typing import TypeAlias


Conversion: TypeAlias = Callable[[str], int | float]


def non_numeric(test: Conversion, samples: Iterable[str]) -> dict[str, int]:
    bad_data = defaultdict(int)
```

```
    for s in samples:
        try:
            test(s)
        except ValueError:
            bad_data[s] += 1
    return bad_data
```

The idea is to apply a conversion function, commonly `int()` or `float()`, but `decimal.Decimal()` may be useful for currency data or other data with a fixed number of decimal places. If the conversion function fails, the exceptional data is preserved in a mapping showing the counts.

You're encouraged to try this with a sequence of strings like the following:

```
data = ["2", "3.14", "42", "Nope", None, ""]
non_numeroc(int, data)
```

This kind of test case will let you see how this function works with good (and bad) data. It can help to transform the test case into a docstring, and include it in the function definition.

If the result of the `non_numeric()` function is an empty dictionary, then the lack of non-numeric data means all of the data is numeric.

The test function is provided first to follow the pattern of higher-order functions like `map()` and `filter()`.

A variation on this function can be used as a numeric filter to pass the numeric values and reject the non-numeric values. This would look like the following:

```
from collections.abc import Iterable, Iterator, Callable
from typing import TypeVar


Num = TypeVar('Num')
```

```
def numeric_filter(
    conversion: Callable[[str], Num],
    samples: Iterable[str]
) -> Iterator[Num]:
        for s in samples:
            try:
                    yield conversion(s)
            except ValueError:
                pass
```

This function will silently reject the values that cannot be converted. The net effect of omitting the data is to create a NULL that does not participate in further computations. An alternative may be to replace invalid values with default values. An even more complicated choice is to interpolate a replacement value using adjacent values. Omitting samples may have a significant impact on the statistical measures used in later stages of processing. This `numeric_filter()` function permits the use of other statistical functions to locate outliers.

For data with good documentation or a data contract, outlier values like -99.99 are easy to spot. For data without good documentation, a statistical test might be more appropriate. See `https://www.itl.nist.gov/div898/handbook/eda/section3/eda35h.htm` for details on approaches to locating outliers.

One approach suitable for small data sets is to use a median-based Z-score. We'll dive into an algorithm that is built on a number of common statistical measures. This will involve computing the median using a function available in the built-in `statistics` package.

For more information on basic statistics for data analytics, see *Statistics for Data Science*.

`https://www.packtpub.com/product/statistics-for-data-science/9781788290678`.

The conventional Z-score for a sample, Z_i, is based on the mean, $\bar{Y}$, and the standard deviation, σ_Y. It's computed as $Z_i = \frac{Y_i - \bar{Y}}{\sigma_Y}$. It measures how many standard deviations a value lies from the mean. Parallel with this is the idea of a median-based Z-score, M_i. The median-based Z-score uses the median, $\tilde{Y}$, and the median absolute deviation, MAD_Y.

This is computed as $M_i = \frac{Y_i - \tilde{Y}}{\text{MAD}_Y}$. This measures how many "MAD" units a value lies from the median of the samples.

The MAD is the median of the absolute values of deviations from the median. It requires computing an overall median, $\tilde{Y}$, then computing all the deviations from the overall median, $Y_i - \tilde{Y}$. From this sequence of deviations from the median, the median value is selected to locate a central value for all of the median absolute deviations. This is computed as $\text{MAD}_Y = \text{median}(|Y_i - \tilde{Y}|)$.

The filter based on M_i looks for any absolute value of the deviation from MAD_Y that's greater than 3.5, $|M_i| > 3.5$. These samples are possible outliers because their absolute deviation from the median is suspiciously large.

To be complete, here's a cell to read the source data:

```python
with series_4_path.open() as source_file:
    series_4_data = list(samples_iter(source_file))
```

This can be followed with a cell to compute the median and the median absolute deviation. The median computation can be done with the `statistics` module. The deviations can then be computed with a generator, and the median absolute deviation computed from the generator. The cell looks like the following:

```python
from statistics import median


y_text = (s['y'] for s in series_4_data)
y = list(numeric_filter(float, y_text))
m_y = median(y)
mad_y = median(abs(y_i - m_y) for y_i in y)
outliers_y = list(
    filter(lambda m_i: m_i > 3.5, ((y_i - m_y)/mad_y for y_i in y))
)
```

The value of `y_text` is a generator that will extract the values mapped to the `'y'` key in each of the raw data samples in the NDJSON file. From these text values, the value of y is computed by applying the `numeric_filter()` function.

It's sometimes helpful to show that `len(y) == len(y_text)` to demonstrate that all values are numeric. In some data sets, the presence of non-numeric data might be a warning that there are deeper problems.

The value of `m_y` is the median of the y values. This is used to compute the MAD value as the median of the absolute deviations from the median. This median absolute deviation provides an expected range around the median.

The `outliers_y` computation uses a generator expression to compute the median-based Z-score, and then keep only those scores that are more than 3.5 MADs from the median.

The data in Series IV of Anscombe's Quartet seems to suffer from an even more complicated outlier problem. While the "x" attribute has a potential outlier, the "y" attribute's MAD is zero. This means more than half the "y" attribute values are the same. This single value is the median, and the difference from the median will be zero for most of the samples.

This anomaly would become an interesting part of the notebook.

Dealing with currency and related values

Most currencies around the world use a fixed number of decimal places. The United States, for example, uses exactly two decimal places for money. These are decimal values; the `float` type is almost always the wrong type for these values.

Python has a `decimal` module with a `Decimal` type, which must be used for currency.

Do not use `float` for currency or anything used in currency-related computations.

Tax rates, discount rates, interest rates, and other money-related fields are also decimal values. They're often used with currency values, and computations must be done using decimal arithmetic rules.

When we multiply `Decimal` values together, the results may have additional digits to the right of the decimal place. This requires applying rounding rules to determine how to round or truncate the extra digits. The rules are essential to getting the correct results. The `float` type `round()` function may not do this properly. The `decimal` module includes a wide variety of rounding and truncating algorithms.

Consider an item with a price of $12.99 in a locale that charges a sales tax of 6.25% on each purchase. This is not a tax amount of $0.811875. The tax amount must be rounded; there are many, many rounding rules in common use by accountants. It's essential to know which rule is required to compute the correct result.

Because the underlying assumption behind currency is decimal computation, the `float` should never be used for currency amounts.

This can be a problem when spreadsheet data is involved. Spreadsheet software generally uses `float` values with complex formatting rules to produce correct-looking answers. This can lead to odd-looking values in a CSV extract like 12.999999997 for an attribute that should have currency values.

Additionally, currency may be decorated with currency symbols like $, £, or €. There may also be separator characters thrown in, depending on the locale. For the US locale, this can mean stray "," characters may be present in large numbers.

The ways currency values may have text decoration suggest the conversion function used by a `non_numeric()` or `numeric_filter()` function will have to be somewhat more sophisticated than the simple use of the `Decimal` class.

Because of these kinds of anomalies, data inspection is a critical step in data acquisition and analysis.

Dealing with intervals or durations

Some date will include duration data in the form `"12:34"`, meaning 12 hours and 34 minutes. This looks exactly like a time of day. In some cases, it might have the form `12h 34m`, which is a bit easier to parse. Without metadata to explain if an attribute is a duration or a time of day, this may be impossible to understand.

For durations, it's helpful to represent the values as a single, common time unit. Seconds are a popular choice. Days are another common choice.

We can create a cell with a given string, for example:

```
time_text = "12:34"
```

Given this string, we can create a cell to compute the duration in seconds as follows:

```
import re


m = re.match(r"(\d+):(\d+)", time_text)
h, m = map(int, m.groups())
sec = (h*60 + m) * 60
sec
```

This will compute a duration, `sec`, of 45,240 seconds from the source time as text, `time_text`. The final expression `sec` in a Jupyter notebook cell will display this variable's value to confirm the computation worked. This cardinal value computation works out elegantly.

For formatting purposes, the inverse computation can be helpful. A floating-point value like 45,240 can be converted back into a sequence of integers, like (12, 34, 0), which can be formatted as "12:34" or "12h 34m 0s".

It might look like this:

```
h_m, s = divmod(sec, 60)
h, m = divmod(h_m, 60)
text = f"{h:02d}:{m:02d}"
text
```

This will produce the string 12:34 from the value of seconds given in the `sec` variable. The final expression `text` in a cell will display the computed value to help confirm the cell works.

It's important to normalize duration strings and complex-looking times into a single float value.

Now that we've looked at some of the tricky cardinal data fields, we can look at the notebook as a whole. In the next section, we'll look at refactoring the notebook to create a useful module.

Extract notebook functions

The computation of ordinary Z-scores and median-based Z-scores are similar in several ways. Here are some common features we might want to extract:

- Extracting the center and variance. This might be the mean and standard deviation, using the `statistics` module. Or it might be the median and MAD.

- Creating a function to compute Z-scores from the mean or median.

- Applying the `filter()` function to locate outliers.

When looking at data with a large number of attributes, or looking at a large number of related data sets, it's helpful to write these functions first in the notebook. Once they've been debugged, they can be cut from the notebook and collected into a separate module. The notebook can then be modified to import the functions, making it easier to reuse these functions.

Because the source data is pushed into a dictionary with string keys, it becomes possible to consider functions that work across a sequence of key values. We might have cells that look like the following example:

```
for column in ('x', 'y'):
    values = list(
        numeric_filter(float, (s[column] for s in series_4_data))
    )
    m = median(values)
    print(column, len(series_4_data), len(values), m)
```

This will analyze all of the columns named in the surrounding `for` statement. In this example, the x and y column names are provided as the collection of columns to analyze. The result is a small table of values with the column name, the raw data size, the filtered data size, and the median of the filtered data.

The idea of a collection of descriptive statistics suggests a class to hold these. We might add the following dataclass:

```
from dataclasses import dataclass


@dataclass
class AttrSummary:
    name: str
    raw_count: int
    valid_count: int
    median: float


    @classmethod
    def from_raw(
            cls: Type["AttrSummary"],
            column: str,
            text_values: list[str]
    ) -> "AttrSummary":
        values = list(numeric_filter(float, text_values))
        return cls(
            name=column,
            raw_count=len(text_values),
            valid_count=len(values),
            median=median(values)
        )
```

The class definition includes a class method to build instances of this class from a collection of raw values. Putting the instance builder into the class definition makes it slightly easier to add additional inspection attributes and the functions needed to compute those attributes. A function that builds `AttrSummary` instances can be used to summarize the attributes of a data set. This function might look like the following:

```python
from collections.abc import Iterator
from typing import TypeAlias


Samples: TypeAlias = list[dict[str, str]]


def summary_iter(
        samples: Samples,
        columns: list[str]
) -> Iterator[AttrSummary]:
    for column in columns:
        text = [s[column] for s in samples]
        yield AttrSummary.from_raw(column, text)
```

This kind of function makes it possible to reuse inspection code for a number of attributes in a complicated data set. After looking at the suggested technical approach, we'll turn to the deliverables for this project.

Deliverables

This project has the following deliverables:

- A `requirements-dev.txt` file that identifies the tools used, usually `jupyterlab==3.5.3`.

- Documentation in the `docs` folder.

- Unit tests for any new changes to the modules in use.

- Any new application modules with code to be used by the inspection notebook.

- A notebook to inspect the attributes that appear to have cardinal data.

This project will require a notebooks directory. See *List of deliverables* for some more information on this structure.

We'll look at a few of these deliverables in a little more detail.

Inspection module

You are encouraged to refactor functions like samples_iter(), non_numeric(), and numeric_filter() into a separate module. Additionally, the AttrSummary class and the closely related summary_iter() function are also good candidates for being moved to a separate module with useful inspection classes and functions.

Notebooks can be refactored to import these classes and functions from a separate module.

It's easiest to throw this module into the notebooks folder to make it easier to access. An alternative is to include the src directory on the PYTHONPATH environment variable, making it available to the Jupyter Lab session.

Another alternative is to create an IPython profile with the ipython profile create command at the terminal prompt. This will create a ~/.ipython/profile_default directory with the default configuration files in it. Adding a startup folder permits including scripts that will add the src directory to the sys.path list of places to look for modules.

See https://ipython.readthedocs.io/en/stable/interactive/tutorial.html#startup-files.

Unit test cases for the module

The various functions were refactored from a notebook to create a separate module need unit tests. In many cases, the functions will have doctest examples; the notebook as a whole will have a doctest cell.

In this case, an extra option in the **pytest** command will execute these tests, as well.

```
% pytest --doctest-modules notebooks/*.py
```

The `--doctest-modules` option will look for the doctest examples and execute them.

An alternative is to use the Python `doctest` command directly.

```
% python -m doctest notebooks/*.py
```

It is, of course, essential to test the code extracted from the notebook to be sure it works properly and can be trusted.

This revised and expanded inspection notebook lets an analyst inspect unknown data sources to confirm values are likely to be cardinal numbers, for example, measures or counts. Using a filter function can help locate invalid or other anomalous text. Some statistical techniques can help to locate outlying values.

In the next project, we'll look at non-cardinal data. This includes nominal data (i.e., strings of digits that aren't numbers), and ordinal values that represent ranking or ordering positions.

Project 2.3: Validating text and codes — nominal data and ordinal numbers

Description

In the previous project (*Project 2.2: Validating cardinal domains — measures, counts, and durations*), we looked at attributes that contained cardinal data – measures and counts. We also need to look at ordinal and nominal data. Ordinal data is generally used to provide ranks and ordering. Nominal data is best thought of as codes made up of strings of digits. Values like US postal codes and bank account numbers are nominal data.

When we look at the **CO_2 PPM — Trends in Atmospheric Carbon Dioxide** data set, available at `https://datahub.io/core/co2-ppm`, it has dates that are provided in two forms: as a `year-month-day` string and as a decimal number. The decimal number positions the first day of the month within the year as a whole.

It's instructive to use ordinal day numbers to compute unique values for each date and compare these with the supplied "Decimal Date" value. An integer day number may be more useful than the decimal date value because it avoids truncation to three decimal places.

Similarly, many of the data sets available from `https://berkeleyearth.org/data/` contain complicated date and time values. Looking at the source data, `https://berk eleyearth.org/archive/source-files/` has data sets with nominal values to encode precipitation types or other details of historical weather. For even more data, see `https://www.ncdc.noaa.gov/cdo-web/`. All of these datasets have dates in a variety of formats.

What's important for this project is to get an overview of the data that involves dates and nominal code values. Later projects will look at cleaning and converting the data for further use. This notebook is only designed to preview and inspect the data. It is used to demonstrate the data is complete and consistent and can be used for further processing.

Dates and times

A date, time, and the combined date-time value represent a specific point in time, sometimes called a timestamp. Generally, these are modeled by Python `datetime` objects.

A date in isolation can generally be treated as a `datetime` with a time of midnight. A time in isolation is often part of a date stated elsewhere in the data or assumed from context. Ideally, a date-time value has been broken into separate columns of data for no good reason and can be combined. In other cases, the data might be a bit more difficult to track down. For example, a log file as a whole might have an implied date — because each log file starts at midnight UTC — and the time values must be combined with the (implied) log's date.

Date-time values are quite complex and rich with strange quirks. To keep the Gregorian calendar aligned with the positions of stars, and the Moon, leap days are added periodically. The `datetime` library in Python is the best way to work with the calendar.

It's generally a bad idea to do any date-time computation outside the `datetime` package.

Home-brewed date computations are difficult to implement correctly.

The `toordinal()` function of a `datetime.datetime` object provides a clear relationship between dates and an ordinal number that can be used to put dates into order.

Because months are irregular, there are several common kinds of date computations:

- A date plus or minus a duration given in months. The day of the month is generally preserved, except in the unusual case of February 29, 30, or 31, where ad hoc rules will apply.

- A date plus or minus a duration given in days or weeks.

These kinds of computations can result in dates in a different year. For month-based computations, an ordinal month value needs to be computed from the date. Given a date, d, with a year, $d.y$, and a month $d.m$, the ordinal month, m_o, is $d.y \times 12 + d.m - 1$. After a computation, the `divmod()` function will recover the year and month of the result. Note that months are generally numbered from 1, but the ordinal month computation numbers months from zero. This leads to a -1 when creating an ordinal month from a date, and a $+1$ when creating a date from an ordinal month. As noted above, when the resulting month is February, something needs to be done to handle the exceptional case of trying to build a possibly invalid date with a day number that's invalid in February of the given year.

For day- or week-based computations, the `toordinal()` function and `fromordinal()` functions will work correctly to order and compute differences between dates.

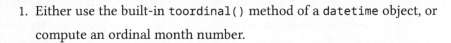

All calendar computations must be done using ordinal values.

Here are the three steps:

1. Either use the built-in `toordinal()` method of a `datetime` object, or compute an ordinal month number.

2. Apply duration offsets to the ordinal value.

3. Either use the built-in `fromordinal()` class method of the `datetime` class, or use the `divmod()` function to compute the year and month of the ordinal month number.

For some developers, the use of ordinal numbers for dates can feel complicated. Using `if` statements to decide if an offset from a date is in a different year is less reliable and requires more extensive edge-case testing. Using an expression like `year, month = divmod(date, 12)` is much easier to test.

In the next section, we'll look at time and the problem of local time.

Time values, local time, and UTC time

Local time is subject to a great deal of complex-seeming rules, particularly in the US. Some countries have a single time zone, simplifying what constitutes local time. In the US, however, each county decides which timezone it belongs to, leading to very complex situations that don't necessarily follow US state borders.

Some countries (the US and Europe, as well as a scattering of other places) offset the time (generally, but not universally by one hour) for part of the year. The rules are not necessarily nationwide; Canada, Mexico, Australia, and Chile have regions that don't have daylight savings time offsets. The Navajo nation — surrounded by the state of Arizona in the US — doesn't switch its clocks.

The rules are here: `https://data.iana.org/time-zones/tz-link.html`. This is part of the Python `datetime` library and is already available in Python.

This complexity makes use of the **universal coordinated time (UTC)** imperative.

> Local times should be converted into UTC for analysis purposes.
>
> See `https://www.rfc-editor.org/rfc/rfc3339` for time formats that can include a local-time offset.
>
> UTC can be converted back into local time to be displayed to users.

Approach

Dates and times often have bewildering formats. This is particularly true in the US, where dates are often written as numbers in month/day/year format. Using year/month/day puts the values in order of significance. Using day/month/year is the reverse order of significance. The US ordering is simply strange.

This makes it difficult to do inspections on completely unknown data without any metadata to explain the serialization format. A date like 01/02/03 could mean almost anything.

In some cases, a survey of many date-like values will reveal a field with a range of 1-12 and another field with a range of 1-31, permitting analysts to distinguish between the month and day. The remaining field can be taken as a truncated year.

In cases where there is not enough data to make a positive identification of month or day, other clues will be needed. Ideally, there's metadata to define the date format.

The `datetime.strptime()` function can be used to parse dates when the format(s) are known. Until the date format is known, the data must be used cautiously.

Here are two Python modules that can help parse dates:

- `https://pypi.org/project/dateparser/`

- `https://pypi.org/project/python-dateutil/`

It's important to carefully inspect the results of date parsing to be sure the results are sensible. There are some confounding factors.

Years, for example, can be provided as two or four digits. For example, when dealing with old data, it's important to note the use of two-digit encoding schemes. For a few years prior to 2000, the year of date might have been given as a complicated two-digit transformation. In one scheme, values from 0 to 29 meant years 2000 to 2029. Values from 30 to 99 meant years 1930 to 1999. These rules were generally ad hoc, and different enterprises may have used different year encodings.

Additionally, leap seconds have been added to the calendar a few times as a way to keep the clocks aligned with planetary motion. Unlike leap years, these are the result of ongoing research by astronomers, and are not defined by the way leap years are defined.

See `https://www.timeanddate.com/time/leapseconds.html` for more information.

The presence of a leap second means that a timestamp like `1972-06-30T23:59:60` is valid. The 60 value for seconds represents the additional leap second. As of this book's initial publication, there were 26 leap seconds, all added on June 30 or December 31 of a given year. These values are rare but valid.

Nominal data

Nominal data is not numeric but may consist of strings of digits, leading to possible sources of confusion and — in some cases — useless data conversions. While nominal data should be treated as text, it's possible for a spreadsheet to treat US Postal ZIP codes as numbers and truncate the leading zeroes. For example, North Adams, MA, has a ZIP code of 01247. A spreadsheet might lose the leading zero, making the code 1247.

While it's generally best to treat nominal data as text, it may be necessary to reformat ZIP codes, account numbers, or part numbers to restore the leading zeroes. This can be done in a number of ways; perhaps the best is to use f-strings to pad values on the left with leading "0" characters. An expression like `f"{zip:0>5s}"` creates a string from the `zip` value using a format of `0>5s`. This format has a padding character, `0`, a padding rule of `>`, and a target size of 5. The final character `s` is the type of data expected; in this case, a string.

An alternative is something like `(5*"0" + zip)[-5:]` to pad a given `zip` value to 5 positions. This prepends zeroes and then takes the right-most five characters. It doesn't seem as elegant as an f-string but can be more flexible.

Extend the data inspection module

In the previous project, *Project 2.2: Validating cardinal domains — measures, counts, and durations*, we considered adding a module with some useful functions to examine cardinal data. We can also add functions for ordinal and nominal data.

For a given problem domain, the date parsing can be defined as a separate, small function. This can help to avoid the complicated-looking `strptime()` function. In many cases, there are only a few date formats, and a parsing function can try the alternatives. It might look like this:

```python
import datetime


def parse_date(source: str) -> datetime.datetime:
    formats = "%Y-%m-%d", "%y-%m-%d", "%Y-%b-%d"
    for fmt in formats:
        try:
            return datetime.datetime.strptime(source, fmt)
        except ValueError:
            pass
    raise ValueError(f"datetime data {source!r} not in any of {formats}
        format")
```

This function has three date formats that it attempts to use to convert the data. If none of the formats match the data, a `ValueError` exception is raised.

For rank ordering data and codes, a notebook cell can rely on a `collections.Counter` instance to get the domain of values. More sophisticated processing is not required for simple numbers and nominal codes.

Deliverables

This project has the following deliverables:

- A `requirements-dev.txt` file that identifies the tools used, usually `jupyterlab==3.5.3`.

- Documentation in the `docs` folder.

- Unit tests for any new changes to the modules in use.

- Any new application modules with code to be used by the inspection notebook.

- A notebook to inspect the attributes that appear to have ordinal or nominal data.

The project directory structure suggested in *Chapter 1, Project Zero: A Template for Other Projects* mentions a `notebooks` directory. See *List of deliverables* for some more information.

For this project, the notebook directory is needed.

We'll look at a few of these deliverables in a little more detail.

Revised inspection module

Functions for date conversions and cleaning up nominal data can be written in a separate module. Or they can be developed in a notebook, and then moved to the inspection module. As we noted in the *Description* section, this project's objective is to support the inspection of the data and the identification of special cases, data anomalies, and outlier values.

Later, we can look at refactoring these functions into a more formal and complete data cleansing module. This project's goal is to inspect the data and write some useful functions for the inspection process. This will create seeds to grow a more complete solution.

Unit test cases

Date parsing is — perhaps — one of the more awkwardly complicated problems. While we often think we've seen all of the source data formats, some small changes to upstream applications can lead to unexpected changes for data analysis purposes.

Every time there's a new date format, it becomes necessary to expand the unit tests with the bad data, and then adjust the parser to handle the bad data. This can lead to a surprisingly large number of date-time examples.

When confronted with a number of very similar cases, the `pytest` parameterized fixtures are very handy. These fixtures provide a number of examples of a test case.

The fixture might look like the following:

```
import pytest


EXAMPLES = [
    ('2021-01-18', datetime.datetime(2021, 1, 18, 0, 0)),
    ('21-01-18', datetime.datetime(2021, 1, 18, 0, 0)),
    ('2021-jan-18', datetime.datetime(2021, 1, 18, 0, 0)),
]
```

```
@pytest.fixture(params=EXAMPLES)
def date_example(request):
    return request.param
```

Each of the example values is a two-tuple with input text and the expected `datetime` object. This pair of values can be decomposed by the test case.

A test that uses this fixture full of examples might look like the following:

```
def test_date(date_example):
    text, expected = date_example
    assert parse_date(text) == expected
```

This kind of test structure permits us to add new formats as they are discovered. The test cases in the `EXAMPLES` variable are easy to expand with additional formats and special cases.

Now that we've looked at inspecting cardinal, ordinal, and nominal data, we can turn to a more specialized form of nominal data: key values used to follow references between separate data sets.

Project 2.4: Finding reference domains

In many cases, data is decomposed to avoid repetition. In *Chapter 5, Data Acquisition Features: SQL Database*, we touched on the idea of normalization to decompose data.

As an example, consider the data sets in this directory: `https://www.ncei.noaa.gov/pu b/data/paleo/historical/northamerica/usa/new-england/`

There are three separate files. Here's what we see when we visit the web page.

Here's the index of the `/pub/data/paleo/historical/northamerica/usa/new-england` file:

Name	Last modified	Size	Description
Parent Directory		-	
new-england-oldweather-data.txt	2014-01-30 13:02	21M	
readme-new-england-oldweather.txt	2014-01-29 19:22	9.6K	
town-summary.txt	2014-01-29 18:51	34K	

The `readme-new-england-oldweather.txt` file has descriptions of a number of codes and their meanings used in the main data set. The "readme" file provides a number of mappings from keys to values. The keys are used in the massive "oldweather-data" file to reduce the repetition of data.

These mappings include the following:

- The Temperature Code Key

- The Precipitation Type Key

- The Precipitation Amount key

- The Snowfall Amount key

- The Like Values Code key

- The Pressure Code Key

- The Sky Cover Key

- The Sky Classification Key

- The Location Code Key

This is a rather complex decomposition of primary data into coded values.

Description

In cases where data is decomposed or normalized, we need to confirm that references between items are valid. Relationships are often one-way — a sample will have a reference to an item in another collection of data. For example, a climate record may have a reference to "Town Id" (TWID) with a value like NY26. A second data set with the "location code key" provides detailed information on the definition of the NY26 town ID. There's no reverse reference from the location code data set to all of the climate records for that location.

We often depict this relationship as an ERD. For example, *Figure 7.1.*

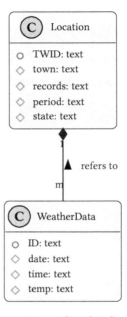

Figure 7.1: A Normalized Relationship

A number of weather data records refer to a single location definition.

The database designers will call the Location's "TWID" attribute a **primary key**. The WeatherData's ID attribute is called a **foreign key**; it's a primary key for a different class of entities. These are often abbreviated as PK and FK.

There are two closely related questions about the relationship between entities:

- What is the *cardinality* of the relationship? This must be viewed from both directions. How many primary key entities have relationships with foreign key entities? How many foreign key entities have a relationship with a primary key entity?

- What is the *optionality* of the relationship? Again, we must ask this in both directions.

Must a primary entity have any foreign key references? Must the foreign key item have a primary key reference?

While a large number of combinations are possible, there are a few common patterns.

- The mandatory many-to-one relationship. This is exemplified by the historical weather data. Many weather data records must refer to a single location definition. There are two common variants. In one case, a location **must** have one or more weather records. The other common variant may have locations without any weather data that refers to the location.

- An optional one-to-one relationship. This isn't in the weather data example, but we may have invoices with payments and invoices without payments. The relationship is one-to-one, but a payment may not exist yet.

- A many-to-many relationship. An example of a many-to-many relationship is a product entity that has a number of features. Features are reused between products. This requires a separate many-to-many association table to track the relationship links.

This leads to the following two detailed inspections:

1. The domain of primary key values. For example, the "TWID" attribute of each location.

2. The domain of the foreign key values. For example, the ID attribute of each weather data record.

If these two sets are identical, we can be sure the foreign keys all have matching primary keys. We can count the number of rows that share a foreign key to work out the cardinality (and the optionality) of the relationship.

If the two sets are not identical, we have to determine which set has the extra rows. Let's call the two sets P and F. Further, we know that $P \neq F$. There are a number of scenarios:

- $P \supset F$: This means there are some primary keys without any foreign keys. If the relationship is optional, then, there's no problem. The $P \setminus F$ is the set of unused entities.

- $F \subset P$: This means there are foreign keys that do not have an associated primary key. This situation may be a misunderstanding of the key attributes, or it may mean data is missing.

What's important for this project is to have an overview of key values and their relationship. This notebook is only designed to preview and inspect the data. It is used to demonstrate the data is complete and consistent, and can be used for further processing.

In the next section, we'll look at how we can build cells in a notebook to compare the keys and determine the cardinality of the relationships.

Approach

To work with data sets like `https://www.ncei.noaa.gov/pub/data/paleo/historical/northamerica/usa/new-england/` we'll need to compare keys.

This will lead to two kinds of data summarization cells in an inspection notebook:

- Summarizing primary keys in a `Counter` object.

- Summarizing foreign key references to those primary keys, also using a `Counter`.

Once the `Counter` summaries are available, then the `.keys()` method will have the distinct primary or foreign key values. This can be transformed into a Python `set` object, permitting elegant comparison, subset checking, and set subtraction operations.

We'll look at the programming to collect key values and references to keys first. Then, we'll look at summaries that are helpful.

Collect and compare keys

The core inspection tool is the `collections.Counter` class. Let's assume we have done two separate data acquisition steps. The first extracted the location definitions from the `readme-new-england-oldweather.txt` file.

The second converted all of the new-england-oldweather-data.txt weather data records into a separate file.

The inspection notebook can load the location definitions and gather the values of the TWID attribute.

One cell for loading the key definitions might look like this:

```
from pathlib import Path
from inspection import samples_iter


location_path = Path("/path/to/location.ndjson")
with location_path.open() as data_file:
    locations = list(samples_iter(data_file))
```

A cell for inspecting the definitions of the town keys might look like this:

```
import collections


town_id_count = collections.Counter(
    row['TWID'] for row in locations
)
town_id_set = set(town_id_count.keys())
```

This creates the town_id_set variable with a set of IDs in use. The values of the town_id_counts variable are the number of location definitions for each ID. Since this is supposed to be a primary key, it should have only a single instance of each value.

The data with references to the town keys may be much larger than the definitions of the keys. In some cases, it's not practical to load all the data into memory, and instead, the inspection needs to work with summaries of selected columns.

For this example, that means a list object is **not** created with the weather data. Instead, a generator expression is used to extract a relevant column, and this generator is then used to build the final summary Counter object.

The rows of data with references to the foreign keys might look like this:

```
weather_data_path = Path("/path/to/weather-data.ndjson")
with weather_data_path.open() as data_file:
    weather_reports = samples_iter(data_file)
    weather_id_count = collections.Counter(
        row['ID'] for row in weather_reports
```

Once the `weather_id_count` summary has been created, the following cell can compute the domain of key references like this:

```
weather_id_set = set(weather_id_count.keys())
```

It's important to note that this example emphatically does *not* create a list of individual weather report samples. That would be a lot of data jammed into memory at one time. Instead, this example uses a generator expression to extract the `'ID'` attribute from each row. These values are used to populate the `weather_id_count` variable. This is used to extract the set of IDs in use in the weather reports.

Since we have two sets, we can use Python's set operations to compare the two. Ideally, a cell can assert that `weather_id_set == town_id_set`. If the two are not equal, then the set subtraction operation can be used to locate anomalous data.

Summarize keys counts

The first summary is the comparison of primary keys to foreign keys. If the two sets don't match, the list of missing foreign keys may be helpful for locating the root cause of the problem.

Additionally, the range of counts for a foreign key provides some hints as to its cardinality and optionality. When a primary key has no foreign key values referring to it, the relationship appears optional. This should be confirmed by reading the metadata descriptions. The lower and upper bounds on the foreign key counts provide the range of the cardinality. Does this range make sense? Are there any hints in the metadata about the cardinality?

The example data source for this project includes a file with summary counts. The town-summary.txt file has four columns: "STID", "TWID", "YEAR", and "Records". The "STID" is from the location definitions; it's the US state. The "TWID" is the town ID. The "YEAR" is from the weather data; it's the year of the report. Finally, the "Records" attribute is the count of weather reports for a given location and year.

The Town ID and Year form a logical pair of values that can be used to build a collections.Counter object. To fully reproduce this table, though, the location definitions are needed to map a Town ID, "TWID," to the associated state, "STID."

While it's also possible to decompose the "TWID" key to extract the state information from the first two characters, this is not a good design alternative. This composite key is an uncommon kind of key design. It's considerably more common for primary keys to be atomic with no internal information available. A good design treats the key as an opaque identifier and looks up the state information in the associated location definition table from the readme file.

Deliverables

This project has the following deliverables:

- A requirements-dev.txt file that identifies the tools used, usually jupyterlab==3.5.3.

- Documentation in the docs folder.

- Unit tests for any new changes to the modules in use.

- Any new application modules with code to be used by the inspection notebook.

- A notebook to inspect the attributes that appear to have foreign or primary keys.

The project directory structure suggested in *Chapter 1, Project Zero: A Template for Other Projects* mentions a notebooks directory. See *List of deliverables* for some more information. For this project, the notebook directory is needed.

We'll look at a few of these deliverables in a little more detail.

Revised inspection module

The functions for examining primary and foreign keys can be written in a separate module. It's often easiest to develop these in a notebook first. There can be odd discrepancies that arise because of misunderstandings. Once the key examination works, it can be moved to the inspection module. As we noted in the *Description*, this project's objective is to support the inspection of the data and the identification of special cases, data anomalies, and outlier values.

Unit test cases

It's often helpful to create test cases for the most common varieties of key problems: primary keys with no foreign keys and foreign keys with no primary keys. These complications don't often arise with readily available, well-curated data sets; they often arise with enterprise data with incomplete documentation.

This can lead to rather lengthy fixtures that contain two collections of source objects. It doesn't take many rows of data to reveal a missing key; two rows of data are enough to show a key that's present and a row with a missing key.

It's also essential to keep these test cases separate from test cases for cardinal data processing, and ordinal data conversions. Since keys are a kind of nominal data, a key cardinality check may be dependent on a separate function to clean damaged key values.

For example, real data may require a step to add leading zeroes to account numbers before they can be checked against a list of transactions to find transactions for the account. These two operations on account number keys need to be built — and tested — in isolation. The data cleanup application can combine the two functions. For now, they are separate concerns with separate test cases.

Revised notebook to use the refactored inspection model

A failure to resolve foreign keys is a chronic problem in data acquisition applications. This is often due to a wide variety of circumstances, and there's no single process for data inspection. This means a notebook can have a spectrum of information in it. We might see any of the following kinds of cells:

- A cell explaining the sets of keys match, and the data is likely usable.

- A cell explaining some primary keys have no foreign key data. This may include a summary of this subset of samples, separate from samples that have foreign key references.

- A cell explaining some foreign keys that have no primary key. These may may reflect errors in the data. It may reflect a more complex relationship between keys. It may reflect a more complicated data model. It may reflect missing data.

In all cases, an extra cell with some markdown explaining the results is necessary. In the future, you will be grateful because in the past, you left an explanation of an anomaly in your notebook.

Summary

This chapter expanded on the core features of the inspection notebook. We looked at handling cardinal data (measures and counts), ordinal data (dates and ranks), and nominal data (codes like account numbers).

Our primary objective was to get a complete view of the data, prior to formalizing our analysis pipeline. A secondary objective was to leave notes for ourselves on outliers, anomalies, data formatting problems and other complications. A pleasant consequence of this effort is to be able to write some functions that can be used downstream to clean and normalize the data we've found.

Starting in *Chapter 9, Project 3.1: Data Cleaning Base Application*, we'll look at refactoring these inspection functions to create a complete and automated data cleaning and normalization application. That application will be based on the lessons learned while creating inspection notebooks.

In the next chapter, we'll look at one more lesson that's often learned from the initial inspection. We often discover the underlying schema behind multiple, diverse sources of data. We'll look at formalizing the schema definition via JSONSchema, and using the schema to validate data.

Extras

Here are some ideas for you to add to the projects in this chapter.

Markdown cells with dates and data source information

A minor feature of an inspection notebook is some identification of the date, time, and source of the data. It's sometimes clear from the context what the data source is; there may, for example, be an obvious path to the data.

However, in many cases, it's not perfectly clear what file is being inspected or how it was acquired. As a general solution, any processing application should produce a log. In some cases, a metadata file can include the details of the processing steps.

This additional metadata on the source and processing steps can be helpful when reviewing a data inspection notebook or sharing a preliminary inspection of data with others. In many cases, this extra data is pasted into ordinary markdown cells. In other cases, this data may be the result of scanning a log file for key `INFO` lines that summarize processing.

Presentation materials

A common request is to tailor a presentation to users or peers to explain a new source of data, or explain anomalies found in existing data sources. These presentations often involve an online meeting or in-person meeting with some kind of "slide deck" that emphasizes the speaker's points.

Proprietary tools like Keynote or PowerPoint are common for these slide decks.

A better choice is to organize a notebook carefully and export it as `reveal.js` slides.

The RISE extension for Jupyter is popular for this. See `https://rise.readthedocs.io/en /stable/`.

Having a notebook that is **also** the slide presentation for business owners and users provide a great deal of flexibility. Rather than copying and pasting to move data from an inspection notebook to PowerPoint (or Keynote), we only need to make sure each slide has a few key points about the data. If the slide has a data sample, it's only a few rows, which provide supporting evidence for the speaker's remarks.

In many enterprises, these presentations are shared widely. It can be beneficial to make sure the data in the presentation comes directly from the source and is immune to copy-paste errors and omissions.

JupyterBook or Quarto for even more sophisticated output

In some cases, a preliminary inspection of data may involve learning a lot of lessons about the data sources, encoding schemes, missing data, and relationships between data sets. This information often needs to be organized and published.

There are a number of ways to disseminate lessons learned about data:

- Share the notebooks. For some communities of users, the interactive nature of a notebook invites further exploration.

- Export the notebook for publication. One choice is to create a PDF that can be shared. Another choice is to create RST, Markdown, or LaTeX and use a publishing pipeline to build a final, shareable document.

- Use a tool like Jupyter{Book} to formalize the publication of a shareable document.

- Use Quarto to publish a final, shareable document.

For Jupyter{Book}, see `https://jupyterbook.org/en/stable/intro.html`. The larger "Executable{Books}" project (`https://executablebooks.org/en/latest/tools.html`) describes the collection of Python-related tools, including Myst-NB, Sphinx, and some related Sphinx themes. The essential ingredient is using Sphinx to control the final publication.

For Quarto, see `https://quarto.org`. This is somewhat more tightly integrated: it requires a single download of the Quarto CLI. The Quarto tool leverages Pandoc to produce a final, elegant, ready-to-publish file.

You are encouraged to look at ways to elevate the shared notebook to an elegant report that can be widely shared.

8

Project 2.5: Schema and Metadata

It helps to keep the data schema separate from the various applications that share the schema. One way to do this is to have a separate module with class definitions that all of the applications in a suite can share. While this is helpful for a simple project, it can be awkward when sharing data schema more widely. A Python language module is particularly difficult for sharing data outside the Python environment.

This project will define a schema in JSON Schema Notation, first by building `pydantic` class definitions, then by extracting the JSON from the class definition. This will allow you to publish a formal definition of the data being created. The schema can be used by a variety of tools to validate data files and assure that the data is suitable for further analytical use.

The schema is also useful for diagnosing problems with data sources. Validator tools like `jsonschema` can provide detailed error reports that can help identify changes in source data from bug fixes or software updates.

This chapter will cover a number of skills related to data inspection techniques:

- Using the **Pydantic** module for crisp, complete definitions

- Using JSON Schema to create an exportable language-independent definition that anyone can use

- Creating test scenarios to use the formal schema definition

We'll start by looking at the reasons why a formal schema is helpful.

Description

Data validation is a common requirement when moving data between applications. It is extremely helpful to have a clear definition of what constitutes valid data. It helps even more when the definition exists outside a particular programming language or platform.

We can use the JSON Schema (`https://json-schema.org`) to define a schema that applies to the intermediate documents created by the acquisition process. Using JSON Schema enables the confident and reliable use of the JSON data format.

The JSON Schema definition can be shared and reused within separate Python projects and with non-Python environments, as well. It allows us to build data quality checks into the acquisition pipeline to positively affirm the data really fit the requirements for analysis and processing.

Additional metadata provided with a schema often includes the provenance of the data and details on how attribute values are derived. This isn't a formal part of a JSON Schema, but we can add some details to the JSON Schema document that includes provenance and processing descriptions.

The subsequent data cleaning projects should validate the input documents using a source schema. Starting with *Chapter 9, Project 3.1: Data Cleaning Base Application*, the applications should validate their output using the target analytic schema. It can seem silly to have an application both create sample records and also validate those records against a schema. What's important is the schema will be shared, and evolve with the needs of consumers of the data. The data acquisition and cleaning operations, on the other hand, evolve with the data sources. It is all too common for an ad hoc solution to a data problem to seem good but create invalid data.

It rarely creates new problems to validate inputs as well as outputs against a visible, agreed-upon schema. There will be some overhead to the validation operation, but much of the processing cost is dominated by the time to perform input and output, not data validation.

Looking forward to *Chapter 12, Project 3.8: Integrated Data Acquisition Web Service*, we'll see additional uses for a formally-defined schema. We'll also uncover a small problem with using JSON Schema to describe ND JSON documents. For now, we'll focus on the need to use JSON Schema to describe data.

We'll start by adding some modules to make it easier to create JSON Schema documents.

Approach

First, we'll need some additional modules. The `jsonschema` module defines a validator that can be used to confirm a document matches the defined schema.

Additionally, the **Pydantic** module provides a way to create class definitions that can emit JSON Schema definitions, saving us from having to create the schema manually. In most cases, manual schema creation is not terribly difficult. For some cases, though, the schema and the validation rules might be challenging to write directly, and having Python class definitions available can simplify the process.

This needs to be added to the `requirements-dev.txt` file so other developers know to install it.

When using **conda** to manage virtual environments, the command might look like the following:

```
% conda install jsonschema pydantic
```

When using other tools to manage virtual environments, the command might look like the following:

```
% python -m pip install jsonschema pydantic
```

The JSON Schema package requires some supplemental type stubs. These are used by the **mypy** tool to confirm the application is using types consistently. Use the following command to add stubs:

```
% mypy --install-types
```

Additionally, the `pydantic` package includes a **mypy** plug-in that will extend the type-checking capabilities of **mypy**. This will spot more nuanced potential problems with classes defined using `pydantic`.

To enable the plugin, add `pydantic.mypy` to the list of plugins in the **mypy** configuration file, `mypy.ini`. The `mypy.ini` file should look like this:

```
[mypy]
plugins = pydantic.mypy
```

(This file goes in the root of the project directory.)

This plugin is part of the **pydantic** download, and is compatible with **mypy** versions starting with 0.910.

With these two packages, we can define classes with details that can be used to create JSON Schema files. Once we have a JSON Schema file, we can use the schema definition to confirm that sample data is valid.

For more information on **Pydantic**, see `https://docs.pydantic.dev`.

The core concept is to use **Pydantic** to define dataclasses with detailed field definitions. These definitions can be used for data validation in Python. The definition can also be used to emit a JSON Schema document to share with other projects.

The schema definitions are also useful for defining an OpenAPI specification. In *Chapter 12, Project 3.8: Integrated Data Acquisition Web Service*, we'll turn to creating a web service that provides data. The OpenAPI specification for this service will include the schema definitions from this project.

The use of **Pydantic** isn't required. It is, however, very convenient for creating a schema that can be described via JSON Schema. It saves a great deal of fussing with details in JSON syntax.

We'll start with using **Pydantic** to create a useful data model module. This will extend the data models built for projects in earlier chapters.

Define Pydantic classes and emit the JSON Schema

We'll start with two profound modifications to the data model definitions used in earlier chapters. One change is to switch from the `dataclasses` module to the `pydantic.dataclasses` module. Doing this creates the need to explicitly use `dataclasses.field` for individual field definitions. This is generally a small change to an `import` statement to use `from pydantic.dataclasses import dataclass`. The dataclasses `field()` function will need some changes, also, to add additional details used by **pydantic**. The changes should be completely transparent to the existing application; all tests will pass after these changes.

The second change is to add some important metadata to the classes. Where the `dataclasses.field(...)` definition is used, the `metadata={}` attribute can be added to include a dictionary with JSON Schema attributes like the description, title, examples, valid ranges of values, etc. For other fields, the `pydantic.Field()` function must be used to provide a title, description, and other constraints on the field. This will generate a great deal of metadata for us.

See `https://docs.pydantic.dev/usage/schema/#field-customization` for the wide variety of field definition details available.

```python
from pydantic import Field
from pydantic.dataclasses import dataclass


@dataclass
class SeriesSample:
    """
    An individual sample value.
    """

    x: float = Field(title="The x attribute", ge=0.0)
    y: float = Field(title="The y attribute", ge=0.0)


@dataclass
class Series:
    """
    A named series with a collection of values.
    """
    name: str = Field(title="Series name")
    samples: list[SeriesSample] = Field(title="Sequence of samples
      in this series")
```

We've provided several additional details in this model definition module. The details include:

- Docstrings on each class. These will become descriptions in the JSON Schema.

- Fields for each attribute. These, too, become descriptions in the JSON Schema.

- For the x and y attributes of the SeriesSample class definition, we added a ge value. This is a range specification, requiring the values to be greater than or equal to zero.

We've also made extremely profound changes to the model: we've moved from the source data description — which was a number of str values — to the target data description, using float values.

What's central here is that we have two variations on each model:

- **Acquisition**: This is the data as we find it "in the wild." In the examples in this book, some variations of source data are text-only, forcing us to use `str` as a common type. Some data sources will have data in more useful Python objects, permitting types other than `str`.

- **Analysis**: This is the data used for further analysis. These data sets can use native Python objects. For the most part, we'll focus on objects that are easily serialized to JSON. The exception will be date-time values, which don't readily serialize to JSON, but require some additional conversion from a standard ISO text format.

The class examples shown above do not *replace* the `model` module in our applications. They form a second model of more useful data. The recommended approach is to change the initial acquisition model's module name from `model` to `acquisition_model` (or perhaps the shorter `source_model`). This property describes the model with mostly string values as the source. This second model is the `analysis_model`.

The results of the initial investigation into the data can provide narrower and more strict constraints for the analysis model class definitions. See *Chapter 7, Data Inspection Features* for a number of inspections that can help to reveal expected minima and maxima for attribute values.

The **Pydantic** library comes with a large number of customized data types that can be used to describe data values. See `https://docs.pydantic.dev/usage/types/` for documentation. Using the `pydantic` types can be simpler than defining an attribute as a string, and trying to create a regular expression for valid values.

Note that validation of source values isn't central to **Pydantic**. When Python objects are provided, it's entirely possible for the **Pydantic** module to perform a successful data conversion where we might have hoped for an exception to be raised. A concrete example is providing a Python `float` object to a field that requires an `int` value. The `float` object will be converted; an exception will *not* be raised. If this kind of very strict validation of Python objects is required, some additional programming is needed.

In the next section, we'll create a JSON Schema definition of our model. We can either export the definition from the class definition, or we can craft the JSON manually.

Define expected data domains in JSON Schema notation

Once we have the class definition, we can then export a schema that describes the class. Note that the **Pydantic** dataclass is a wrapper around an underlying pydantic.BaseModel subclass definition.

We can create a JSON Schema document by adding the following lines to the bottom of the module:

```python
from pydantic import schema_of
import json


if __name__ == "__main__":
    schema = schema_of(Series)
    print(json.dumps(schema, indent=2))
```

These lines turn the data definition module into a script that writes the JSON Schema definition to the standard output file.

The schema_of() function will extract a schema from the dataclass created in the previous section. (See *Define Pydantic classes and emit the JSON Schema.*) The underlying pydantic.BaseModel subclass also has a schema() method that will transform the class definition into a richly-detailed JSON Schema definition. When working with **pydantic** dataclasses, the pydantic.BaseModel isn't directly available, and the schema_of() function must be used.

When executing the terminal command python src/analysis_model.py, the schema is displayed.

The output begins as follows:

```json
{
  "title": "Series",
```

```
      "description": "A named series with a collection of values.",
      "type": "object",
      "properties": {
        "name": {
          "title": "Series name",
          "type": "string"
        },
        "samples": {
          "title": "Sequence of samples in this series",
          "type": "array",
          "items": {
            "\$ref": "#/definitions/SeriesSample"
          }
        }
      },
      "required": [
        "name",
        "samples"
      ],
      ...
}
```

We can see that the title matches the class name. The description matches the docstring. The collection of properties matches the attributes' names in the class. Each of the property definitions provides the type information from the dataclass.

The $ref item is a reference to another definition provided later in the JSON Schema. This use of references makes sure the other class definition is separately visible, and is available to support this schema definition.

A very complex model may have a number of definitions that are shared in multiple places. This $ref technique normalizes the structure so only a single definition is provided.

Multiple references to the single definition assure proper reuse of the class definition.

The JSON structure may look unusual at first glance, but it's not frighteningly complex. Reviewing `https://json-schema.org` will provide information on how best to create JSON Schema definitions without using the **Pydantic** module.

Use JSON Schema to validate intermediate files

Once we have a JSON Schema definition, we can provide it to other stakeholders to be sure they understand the data required or the data provided. We can also use the JSON Schema to create a validator that can examine a JSON document and determine if the document really does match the schema.

We can do this with a pydantic class definition. There's a `parse_obj()` method that will examine a dictionary to create an instance of the given pydantic class could be built. The `parse_raw()` method can parse a string or bytes object to create an instance of the given class.

We can also do this with the `jsonschema` module. We'll look at this as an alternative to `pydantic` to show how sharing the JSON Schema allows other applications to work with a formal definition of the analysis model.

First, we need to create a validator from the schema. We can dump the JSON into a file and then load the JSON back from the file. We can also save a step by creating a validator directly from the **Pydantic**-created JSON Schema. Here's the short version:

```
from pydantic import schema_of
from jsonschema.validators import Draft202012Validator
from analysis_model import *


schema = schema_of(SeriesSample)
validator = Draft202012Validator(schema)
```

This creates a validator using the latest version of JSON Schema, the 2020 draft. (The project is on track to become a standard, and has gone through a number of drafts as it matures.)

Here's how we might write a function to scan a file to be sure the NDJSON documents all properly fit the defined schema:

```python
def validate_ndjson_file(
        validator: Draft202012Validator,
        source_file: TextIO
) -> Counter[str]:
    counts: Counter[str] = Counter()
    for row in source_file:
        document = json.loads(row)
        if not validator.is_valid(document):
            errors = list(validator.iter_errors(document))
            print(document, errors)
            counts['faulty'] += 1
        else:
            counts['good'] += 1
    return counts
```

This function will read each NDJSON document from the given source file. It will use the given validator to see if the document has problems or is otherwise valid. For faulty documents, it will print the document and the entire list of validation errors.

This kind of function can be embedded into a separate script to check files.

We can, similarly, create the schema for the source model, and use JSON Schema (or **Pydantic**) to validate source files before attempting to process them.

We'll turn to the more complete validation and cleaning solution in *Chapter 9, Project 3.1: Data Cleaning Base Application.* This project is one of the foundational components of the more complete solution.

We'll look at the deliverables for this project in the next section.

Deliverables

This project has the following deliverables:

- A `requirements.txt` file that identifies the tools used, usually `pydantic==1.10.2` and `jsonschema==4.16.0`.

- Documentation in the `docs` folder.

- The JSON-format files with the source and analysis schemas. A separate `schema` directory is the suggested location for these files.

- An acceptance test for the schemas.

We'll look at the schema acceptance test in some detail. Then we'll look at using schema to extend other acceptance tests.

Schema acceptance tests

To know if the schema is useful, it is essential to have acceptance test cases. As new sources of data are integrated into an application, and old sources of data mutate through ordinary bug fixes and upgrades, files will change. The new files will often cause problems, and the root cause of the problem will be the unexpected file format change.

Once a file format change is identified, the smallest relevant example needs to be transformed into an acceptance test. The test will — of course — fail. Now, the data acquisition pipeline can be fixed knowing there is a precise definition of done.

To start with, the acceptance test suite should have an example file that's valid and an example file that's invalid.

As we noted in *Chapter 4, Data Acquisition Features: Web APIs and Scraping*, we can provide a large block of text as part of a Gherkin scenario. We can consider something like the following scenario:

```
Scenario: Valid file is recognized.
    Given a file "example_1.ndjson" with the following content
```

```
"""
{"x": 1.2, "y": 3.4}
{"x": 5.6, "y": 7.8}
"""
When the schema validation tool is run with the analysis schema
Then the output shows 2 good records
And the output shows 0 faulty records
```

This allows us to provide the contents for an NDJSON file. The HTML extract command is quite long. The content is available as the `context.text` parameter of the step definition function. See *Acceptance tests* for more examples of how to write the step definitions to create a temporary file to be used for this test case.

Scenarios for faulty records are also essential, of course. It's important to be sure the schema definition will reject invalid data.

Extended acceptance testing

In *Chapters 3*, *4*, and *5*, we wrote acceptance tests that — generally — looked at log summaries of the application's activity to be sure it properly acquired source data. We did not write acceptance tests that specifically looked at the data.

Testing with a schema definition permits a complete analysis of each and every field and record in a file. The completeness of this check is of tremendous value.

This means that we can add some additional Then steps to existing scenarios. They might look like the following:

```
# Given (shown earlier)...
# When (shown earlier)...
Then the log has an INFO line with "header: ['x', 'y']"
And log has INFO line with "Series_1 count: 11"
And log has INFO line with "Series_2 count: 11"
And log has INFO line with "Series_3 count: 11"
```

```
And log has INFO line with "Series_4 count: 11"
And the output directory files are valid
    using the "schema/Anscombe_Source.json" schema
```

The additional "Then the output directory files are valid…" line requires a step definition that must do the following things:

1. Load the named JSON Schema file and build a `Validator`.

2. Use the `Validator` object to examine each line of the ND JSON file to be sure they're valid.

This use of the schema as part of the acceptance test suite will parallel the way data suppliers and data consumers can use the schema to assure the data files are valid.

It's important to note the schema definition given earlier in this chapter (in *Define Pydantic classes and emit the JSON Schema*) was the output from a future project's data cleaning step. The schema shown in that example is not the output from the previous data acquisition applications.

To validate the output from data acquisition, you will need to use the model for the various data acquisition projects in *Chapters 3, 4*, and *5*. This will be **very** similar to the example shown earlier in this chapter. While similar, it will differ in a profound way: it will use `str` instead of `float` for the series sample attribute values.

Summary

This chapter's projects have shown examples of the following features of a data acquisition application:

- Using the Pydantic module for crisp, complete definitions

- Using JSON Schema to create an exportable language-independent definition that anyone can use

- Creating test scenarios to use the formal schema definition

Having formalized schema definitions permits recording additional details about the data processing applications and the transformations applied to the data.

The docstrings for the class definitions become the descriptions in the schema. This permits writing details on data provenance and transformation that are exposed to all users of the data.

The JSON Schema standard permits recording examples of values. The **Pydantic** package has ways to include this metadata in field definitions, and class configuration objects. This can be helpful when explaining odd or unusual data encodings.

Further, for text fields, JSONSchema permits including a format attribute that can provide a regular expression used to validate the text. The **Pydantic** package has first-class support for this additional validation of text fields.

We'll return to the details of data validation in *Chapter 9, Project 3.1: Data Cleaning Base Application* and *Chapter 10, Data Cleaning Features*. In those chapters, we'll delve more deeply into the various **Pydantic** validation features.

Extras

Here are some ideas for you to add to this project.

Revise all previous chapter models to use Pydantic

The previous chapters used `dataclass` definitions from the `dataclasses` module. These can be shifted to use the `pydantic.dataclasses` module. This should have minimal impact on the previous projects.

We can also shift all of the previous acceptance test suites to use a formal schema definition for the source data.

Use the ORM layer

For SQL extracts, an ORM can be helpful. The pydantic module lets an application create Python objects from intermediate ORM objects. This two-layer processing seems complex but permits detailed validation in the **Pydantic** objects that aren't handled by the database.

For example, a database may have a numeric column without any range provided. A **Pydantic** class definition can provide a field definition with ge and le attributes to define a range. Further, **Pydantic** permits the definition of a unique data type with unique validation rules that can be applied to database extract values.

First, see https://docs.sqlalchemy.org/en/20/orm/ for information on the SQLAlchemy ORM layer. This provides a class definition from which SQL statements like CREATE TABLE, SELECT, and INSERT can be derived.

Then, see the https://docs.pydantic.dev/usage/models/#orm-mode-aka-arbitrary -class-instances "ORM Mode (aka Arbitrary Class Instances)" section of the **Pydantic** documentation for ways to map a more useful class to the intermediate ORM class.

For legacy data in a quirky, poorly-designed database, this can become a problem. For databases designed from the beginning with an ORM layer, on the other hand, this can be a simplification to the SQL.

9

Project 3.1: Data Cleaning Base Application

Data validation, cleaning, converting, and standardizing are steps required to transform raw data acquired from source applications into something that can be used for analytical purposes. Since we started using a small data set of very clean data, we may need to improvise a bit to create some "dirty" raw data. A good alternative is to search for more complicated, raw data.

This chapter will guide you through the design of a data cleaning application, separate from the raw data acquisition. Many details of cleaning, converting, and standardizing will be left for subsequent projects. This initial project creates a foundation that will be extended by adding features. The idea is to prepare for the goal of a complete data pipeline that starts with acquisition and passes the data through a separate cleaning stage. We want to exploit the Linux principle of having applications connected by a shared buffer, often referred to as a shell pipeline.

This chapter will cover a number of skills related to the design of data validation and cleaning applications:

- CLI architecture and how to design a pipeline of processes

- The core concepts of validating, cleaning, converting, and standardizing raw data

We won't address all the aspects of converting and standardizing data in this chapter. Projects in *Chapter 10, Data Cleaning Features* will expand on many conversion topics. The project in *Chapter 12, Project 3.8: Integrated Data Acquisition Web Service* will address the integrated pipeline idea. For now, we want to build an adaptable base application that can be extended to add features.

We'll start with a description of an idealized data cleaning application.

Description

We need to build a data validating, cleaning, and standardizing application. A data inspection notebook is a handy starting point for this design work. The goal is a fully-automated application to reflect the lessons learned from inspecting the data.

A data preparation pipeline has the following conceptual tasks:

- Validate the acquired source text to be sure it's usable and to mark invalid data for remediation.

- Clean any invalid raw data where necessary; this expands the available data in those cases where sensible cleaning can be defined.

- Convert the validated and cleaned source data from text (or bytes) to usable Python objects.

- Where necessary, standardize the code or ranges of source data. The requirements here vary with the problem domain.

The goal is to create clean, standardized data for subsequent analysis. Surprises occur all the time. There are several sources:

- Technical problems with file formats of the upstream software. The intent of the acquisition program is to isolate physical format issues.

- Data representation problems with the source data. The intent of this project is to isolate the validity and standardization of the values.

Once cleaned, the data itself may still contain surprising relationships, trends, or distributions. This is discovered with later projects that create analytic notebooks and reports. Sometimes a surprise comes from finding the *Null Hypothesis* is true and the data only shows insignificant random variation.

In many practical cases, the first three steps — validate, clean, and convert — are often combined into a single function call. For example, when dealing with numeric values, the `int()` or `float()` functions will validate and convert a value, raising an exception for invalid numbers.

In a few edge cases, these steps need to be considered in isolation – often because there's a tangled interaction between validation and cleaning. For example, some data is plagued by dropping the leading zeros from US postal codes. This can be a tangled problem where the data is superficially invalid but can be reliably cleaned before attempting validation. In this case, validating the postal code against an official list of codes comes after cleaning, not before. Since the data will remain as text, there's no actual conversion step after the clean-and-validate composite step.

User experience

The overall **User Experience (UX)** will be two command-line applications. The first application will acquire the raw data, and the second will clean the data. Each has options to fine-tune the `acquire` and `cleanse` steps.

There are several variations on the `acquire` command, shown in earlier chapters. Most notably, *Chapter 3, Project 1.1: Data Acquisition Base Application, Chapter 4, Data Acquisition Features: Web APIs and Scraping*, and *Chapter 5, Data Acquisition Features: SQL Database.*

For the **clean** application, the expected command-line should like something like this:

```
% python src/clean.py -o analysis -i quartet/Series_1.ndjson
```

The `-o analysis` specifies a directory into which the resulting clean data is written.

The `-i quartet/Series_1.ndjson` specifies the path to the source data file. This is a file written by the acquisition application.

Note that we're not using a positional argument to name the input file. The use of a positional argument for a filename is a common provision in many — but not all — Linux commands. The reason to avoid positional arguments is to make it easier to adapt this to become part of a pipeline of processing stages.

Specifically, we'd like the following to work, also:

```
% python src/acquire.py -s Series_1Pair --csv source.csv | \
      python src/clean.py -o analysis/Series_1.ndjson
```

This shell line has two commands, one to do the raw data acquisition, and the other to perform the validation and cleaning. The acquisition command uses the `-s Series_1Pair` argument to name a specific series extraction class. This class will be used to create a single series as output. The `--csv source.csv` argument names the input file to process. Other options could name RESTful APIs or provide a database connection string.

The second command reads the output from the first command and writes this to a file. The file is named by the `-o` argument value in the second command.

This pipeline concept, made available with the shell's | operator, means these two processes will run concurrently. This means data is passed from one process to the other as it becomes available. For very large source files, cleaning data as it's being acquired can reduce processing time dramatically.

In *Project 3.6: Integration to create an acquisition pipeline* we'll expand on this design to include some ideas for concurrent processing.

Now that we've seen an overview of the application's purpose, let's take a look at the source data.

Source data

The earlier projects produced source data in an approximately consistent format. These projects focused on acquiring data that is text. The individual samples were transformed into small JSON-friendly dictionaries, using the NDJSON format. This can simplify the validation and cleaning operation.

The NDJSON file format is described at `http://ndjson.org` and `https://jsonlines.org`.

There are two design principles behind the **acquire** application:

- Preserve the original source data as much as possible.

- Perform the fewest text transformations needed during acquisition.

Preserving the source data makes it slightly easier to locate problems when there are unexpected changes to source applications. Minimizing the text transformations, similarly, keeps the data closer to the source. Moving from a variety of representations to a single representation simplifies the data cleaning and transformation steps.

All of the data acquisition projects involve some kind of textual transformation from a source representation to ND JSON.

Chapter	Section	Source
3	*Chapter 3, Project 1.1: Data Acquisition Base Application*	CSV parsing
4	*Project 1.2: Acquire data from a web service*	Zipped CSV or JSON
4	*Project 1.3: Scrape data from a web page*	HTML
5	*Project 1.5: Acquire data from a SQL extract*	SQL Extract

In some cases — i.e., extracting HTML — the textual changes to peel the markup away from the data is profound. The SQL database extract involves undoing the database's internal representation of numbers or dates and writing the values as text. In some cases, the text transformations are minor.

Result data

The cleaned output files will be ND JSON; similar to the raw input files. We'll address this output file format in detail in *Chapter 11, Project 3.7: Interim Data Persistence*. For this project, it's easiest to stick with writing the JSON representation of a Pydantic dataclass.

For Python's native `dataclasses`, the `dataclasses.asdict()` function will produce a dictionary from a dataclass instance. The `json.dumps()` function will convert this to text in JSON syntax.

For Pydantic dataclasses, however, the `asdict()` function can't be used. There's no built-in method for emitting the JSON representation of a pydantic dataclass instance.

For version 1 of **Pydantic**, a slight change is required to write ND JSON. An expression like the following will emit a JSON serialization of a pydantic dataclass:

```python
import json
from pydantic.json import pydantic_encoder
from typing import TextIO, Iterable


import analysis_model


def persist_samples(
        target_file: TextIO,
        analysis_samples: Iterable[analysis_model.SeriesSample | None]
) -> int:
    count = 0
    for result in filter(None, analysis_samples):
        target_file.write(json.dumps(result, default=pydantic_encoder))
        target_file.write("\n")
        count += 1
    return count
```

This central feature is the `default=pydantic_encoder` argument value for the `json.dumps()` function. This will handle the proper decoding of the dataclass structure into JSON notation.

For version 2 of **pydantic**, there will be a slightly different approach. This makes use of a `RootModel[classname](object)` construct to extract the root model for a given class from an object. In this case, `RootModel[SeriesSample](result).model_dump()` will create a root model that can emit a nested dictionary structure. No special `pydantic_encoder` will be required for version 2.

Now that we've looked at the inputs and outputs, we can survey the processing concepts. Additional processing details will wait for later projects.

Conversions and processing

For this project, we're trying to minimize the processing complications. In the next chapter, *Chapter 10, Data Cleaning Features*, we'll look at a number of additional processing requirements that will add complications. As a teaser for the projects in the next chapter, we'll describe some of the kinds of field-level validation, cleaning, and conversion that may be required.

One example we've focused on, Anscombe's Quartet data, needs to be converted to a series of floating-point values. While this is painfully obvious, we've held off on the conversion from the text to the Python `float` object to illustrate the more general principle of separating the complications of acquiring data from analyzing the data. The output from this application will have each resulting ND JSON document with `float` values instead of `string` values.

The distinction in the JSON documents will be tiny: the use of " for the raw-data strings. This will be omitted for the `float` values.

This tiny detail is important because every data set will have distinct conversion requirements. The data inspection notebook will reveal data domains like text, integers, date-time stamps, durations, and a mixture of more specialized domains. It's essential to examine the data before trusting any schema definition or documentation about the data.

We'll look at three common kinds of complications:

- Fields that must be decomposed.

- Fields which must be composed.

- Unions of sample types in a single collection.

- "Opaque" codes used to replace particularly sensitive information.

One complication is when multiple source values are collapsed into a single field. This single source value will need to be decomposed into multiple values for analysis. With the very clean data sets available from Kaggle, a need for decomposition is unusual. Enterprise data sets, on the other hand, will often have fields that are not properly decomposed into atomic values, reflecting optimizations or legacy processing requirements. For example, a product ID code may include a line of business and a year of introduction as part of the code. For example, a boat's hull ID number might include "421880182," meaning it's a 42-foot hull, serial number 188, completed in January 1982. Three disparate items were all coded as digits. For analytical purposes, it may be necessary to separate the items that comprise the coded value. In other cases, several source fields will need to be combined. An example data set where a timestamp is decomposed into three separate fields can be found when looking at tidal data.

See `https://tidesandcurrents.noaa.gov/tide_predictions.html` for Tide Predictions around the US. This site supports downloads in a variety of formats, as well as RESTful API requests for tide predictions.

Each of the tidal events in an annual tide table has a timestamp. The timestamp is decomposed into three separate fields: the date, the day of the week, and the local time. The day of the week is helpful, but it is entirely derived from the date. The date and time need to be combined into a single datetime value to make this data useful. It's common to use `datetime.combine(date, time)` to merge separate date and time values into a single value.

Sometimes a data set will have records of a variety of subtypes merged into a single collection. The various types are often discriminated from each other by the values of a field. A financial application might include a mixture of invoices and payments; many fields overlap, but the meanings of these two transaction types are dramatically different. A single field with a code value of "I" or "P" may be the only way to distinguish the types of business records represented.

When multiple subtypes are present, the collection can be called a *discriminated union* of subtypes; sometimes simply called a **union**. The discriminator and the subtypes suggest a class hierachy is required to describe the variety of sample types. A common base class is needed to describe the common fields, including the discriminator. Each subclass has a distinct definition for the fields unique to the subclass.

One additional complication stems from data sources with "opaque" data. These are string fields that can be used for equality comparison, but nothing else. These values are often the result of a data analysis approach called **masking**, **deidentification**, or **pseudonymization**. This is sometimes also called "tokenizing" because an opaque token has replaced the sensitive data. In banking, for example, it's common for analytical data to have account numbers or payment card numbers transformed into opaque values. These can be used to aggregate behavior, but cannot be used to identify an individual account holder or payment card. These fields must be treated as strings, and no other processing can be done.

For now, we'll defer the implementation details of these complications to a later chapter. The ideas should inform design decisions for the initial, foundational application.

In addition to clean valid data, the application needs to produce information about the invalid data. Next, we'll look at the logs and error reports.

Error reports

The central feature of this application is to output files with valid, useful data for analytic purposes. We've left off some details of what happens when an acquired document isn't actually usable.

Here are a number of choices related to the observability of invalid data:

- Raise an overall exception and stop. This is appropriate when working with carefully-curated data sets like the Anscombe Quartet.

- Make all of the bad data observable, either through the log or by writing bad data to a separate rejected samples file.

- Silently reject the bad data. This is often used with large data sources where there is no curation or quality control over the source.

In all cases, the summary counts of acquired data, usable analytic data, and cleaned, and rejected data are essential. It's imperative to be sure the number of raw records read is accounted for, and the provenance of cleaned and rejected data is clear. The summary counts, in many cases, are the primary way to observe changes in data sources. A non-zero error count, may be so important that it's used as the final exit status code for the cleaning application.

In addition to the observability of bad data, we may be able to clean the source data. There are several choices here, also:

- Log the details of each object where cleaning is done. This is often used with data coming from a spreadsheet where the unexpected data may be rows that need to be corrected manually.

- Count the number of items cleaned without the supporting details. This is often used with large data sources where changes are frequent.

- Quietly clean the bad data as an expected, normal operational step. This is often used when raw data comes directly from measurement devices in unreliable environments, perhaps in space, or at the bottom of the ocean.

Further, each field may have distinct rules for whether or not cleaning bad data is a significant concern or a common, expected operation. The intersection of observability and automated cleaning has a large number of alternatives.

The solutions to data cleaning and standardization are often a matter of deep, ongoing conversations with users. Each data acquisition pipeline is unique with regard to error reporting and data cleaning.

It's sometimes necessary to have a command-line option to choose between logging each error or simply summarizing the number of errors. Additionally, the application might return a non-zero exit code when any bad records are found; this permits a parent application (e.g., a shell script) to stop processing in the presence of errors.

We've looked at the overall processing, the source files, the result files, and some of the error-reporting alternatives that might be used. In the next section, we'll look at some design approaches we can use to implement this application.

Approach

We'll take some guidance from the C4 model (`https://c4model.com`) when looking at our approach.

- **Context**: For this project, the context diagram has expanded to three use cases: acquire, inspect, and clean.

- **Containers**: There's one container for the various applications: the user's personal computer.

- **Components**: There are two significantly different collections of software components: the acquisition program and the cleaning program.

- **Code**: We'll touch on this to provide some suggested directions.

A context diagram for this application is shown in *Figure 9.1*.

A component diagram for the conversion application isn't going to be as complicated as the component diagrams for acquisition applications. One reason for this is there are no choices for reading, extracting, or downloading raw data files. The source files are the ND JSON files created by the acquisition application.

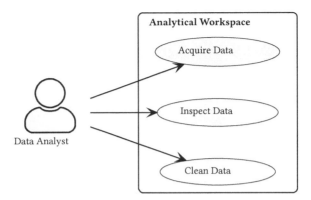

Figure 9.1: Context Diagram

The second reason the conversion programs tend to be simpler is they often rely on built-in Python-type definitions, and packages like pydantic to provide the needed conversion processing. The complications of parsing HTML or XML sources were isolated in the acquisition layer, permitting this application to focus on the problem domain data types and relationships.

The components for this application are shown in *Figure* 9.2.

Note that we've used a dotted "depends on" arrow. This does not show the data flow from acquire to clean. It shows how the clean application depends on the acquire application's output.

The design for the **clean** application often involves an almost purely functional design. Class definitions — of course — can be used. Classes don't seem to be helpful when the application processing involves stateless, immutable objects.

In rare cases, a cleaning application will be required to perform dramatic reorganizations of data. It may be necessary to accumulate details from a variety of transactions, updating the state of a composite object. For example, there may be multiple payments for an invoice that must be combined for reconciliation purposes. In this kind of application, associating payments and invoices may require working through sophisticated matching rules.

Note that the **clean** application and the **acquire** application will both share a common

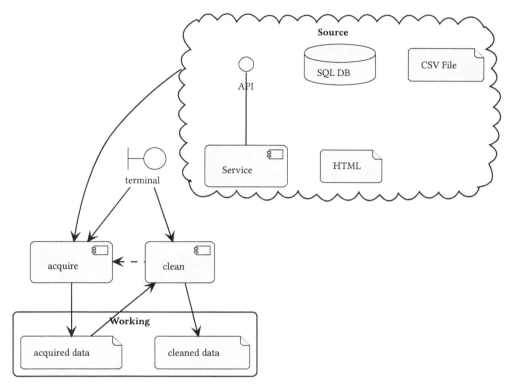

Figure 9.2: Component Diagram

set of dataclasses. These classes represent the source data, the output from the **acquire** application. They also define the input to the **clean** application. A separate set of dataclasses represent the working values used for later analysis applications.

Our goal is to create three modules:

- clean.py: The main application.

- analytical_model.py: A module with dataclass definitions for the pure-Python objects that we'll be working with. These classes will — generally — be created from JSON-friendly dictionaries with string values.

- conversions.py: A module with any specialized validation, cleaning, and conversion functions.

If needed, any application-specific conversion functions may be required to transform source values to "clean," usable Python objects. If this can't be done, the function can instead raise `ValueError` exceptions for invalid data, following the established pattern for functions like Python's built-in `float()` function. Additionally, `TypeError` exceptions may be helpful when the object — as a whole — is invalid. In some cases, the `assert` statement is used, and an `AssertionError` may be raised to indicate invalid data.

For this baseline application, we'll stick to the simpler and more common design pattern. We'll look at individual functions that combine validation and cleaning.

Model module refactoring

We appear to have two distinct models: the "as-acquired" model with text fields, and the "to-be-analyzed" model with proper Python types, like `float` and `int`. The presence of multiple variations on the model means we either need a lot of distinct class names, or two distinct modules as namespaces to keep the classes organized.

The cleaning application is the only application where the acquire and analysis models are **both** used. All other applications either acquire raw data or work with clean analysis data.

The previous examples had a single `model.py` module with dataclasses for the acquired data. At this point, it has become more clear that this was not a great long-term decision. Because there are two distinct variations on the data model, the generic `model` module name needs to be refactored. To distinguish the acquired data model from the analytic data model, a prefix should be adequate: the module names can be `acquire_model` and `analysis_model`.

(The English parts of speech don't match exactly. We'd rather not have to type "acquisition_model". The slightly shorter name seems easier to work with and clear enough.)

Within these two model files, the class names can be the same. We might have names `acquire_model.SeriesSample` and `analysis_model.SeriesSample` as distinct classes.

To an extent, we can sometimes copy the acquired model module to create the analysis model module. We'd need to change `from dataclasses import dataclass` to the **Pydantic** version, `from pydantic import dataclass`. This is a very small change, which makes it easy to start with. In some older versions of **Pydantic** and **mypy**, the **Pydantic** version of `dataclass` doesn't expose the attribute types in a way that is transparent to the **mypy** tool.

In many cases, it can work out well to import `BaseModel` and use this as the parent class for the analytic models. Using the `pydantic.BaseModel` parent class often has a better coexistence with the **mypy** tool. This requires a larger change when upgrading from dataclasses to leverage the **pydantic** package. Since it's beneficial when using the **mypy** tool, it's the path we recommend following.

This **Pydantic** version of `dataclass` introduces a separate validator method that will be used (automatically) to process fields. For simple class definitions with a relatively clear mapping from the acquire class to the analysis class, a small change is required to the class definition.

One common design pattern for this new analysis model class is shown in the following example for **Pydantic** version 1:

```python
from pydantic import validator, BaseModel, Field


class SeriesSample(BaseModel):
    """
    An individual sample value.
    """

    x: float = Field(title="The x attribute", ge=0.0)
    y: float = Field(title="The y attribute", ge=0.0)

    @validator('x', 'y', pre=True)
    def clean_value(cls, value: str | float) -> str:
        match value:
```

```
        case str():
            for char in "\N{ZERO WIDTH SPACE}":
                value = value.replace(char, "")
            return value
        case float():
            return value
```

This design defines a class-level method, `clean_value()`, to handle cleaning the source data when it's a string. The validator has the `@validator()` decorator to provide the attribute names to which this function applies, as well as the specific stage in the sequence of operations. In this case, `pre=True` means this validation applies **before** the individual fields are validated and converted to useful types.

This will be replaced by a number of much more flexible alternatives in **Pydantic** version 2. The newer release will step away from the `pre=True` syntax used to assure this is done prior to the built-in handler accessing the field.

The Pydantic 2 release will introduce a radically new approach using annotations to specify validation rules. It will also retain a decorator that's very similar to the old version 1 validation.

One migration path is to replace `validator` with `field_validator`. This will require changing the `pre=True` or `post=True` with a more universal `mode='before'` or `mode='after'`. This new approach permits writing field validators that "wrap" the conversion handler with both before and after processing.

To use **Pydantic** version two, use `@field_validator('x', 'y', mode='before')` to replace the `@validator` decorator in the example. The `import` must also change to reflect the new name of the decorator.

This validator function handles the case where the string version of source data can include Unicode `U+200B`, a special character called the zero-width space. In Python, we can use `"\N{ZERO WIDTH SPACE}"` to make this character visible. While lengthy, this name seems better than the obscure `"\u200b"`.

(See `https://www.fileformat.info/info/unicode/char/200b/index.htm` for details of this character.)

When a function works in the `pre=True` or `mode='before'` phase, then **pydantic** will automatically apply the final conversion function to complete the essential work of validation and conversion. This additional validator function can be designed, then, to focus narrowly only on cleaning the raw data.

The idea of a validator function must reflect two separate use cases for this class:

1. Cleaning and converting acquired data, generally strings, to more useful analytical data types.

2. Loading already cleaned analytical data, where type conversion is not required.

Our primary interest at this time is in the first use case, cleaning and conversion. Later, starting in chapter *Chapter 13, Project 4.1: Visual Analysis Techniques* we'll switch over to the second case, loading clean data.

These two use cases are reflected in the type hint for the validator function. The parameter is defined as `value: str | float`. The first use case, conversion, expects a value of type `str`. The second use case, loading cleaned data, expects a cleaned value of type `float`. This kind of type of union is helpful with validator functions.

Instances of the analytic model will be built from `acquire_model` objects. Because the acquired model uses `dataclasses`, we can leverage the `dataclasses.asdict()` function to transform a source object into a dictionary. This can be used to perform Pydantic validation and conversion to create the analytic model objects.

We can add the following method in the dataclass definition:

```
@classmethod
def from_acquire_dataclass(
        cls,
        acquired: acquire_model.SeriesSample
```

```
) -> "SeriesSample":
    return SeriesSample(**asdict(acquired))
```

This method extracts a dictionary from the acquired data model's version of the
`SeriesSample` class and uses it to create an instance of the analytic model's variation
of this class. This method pushes all of the validation and conversion work to the **Pydantic**
declarations. This method also requires `from dataclasses import asdict` to introduce
the needed `asdict()` function.

In cases where the field names don't match, or some other transformation is required, a
more complicated dictionary builder can replace the `asdict(acquired)` processing. We'll
see examples of this in *Chapter 10, Data Cleaning Features*, where acquired fields need to
be combined before they can be converted.

We'll revisit some aspects of this design decision in *Chapter 11, Project 3.7: Interim Data
Persistence*. First, however, we'll look at **pydantic** version 2 validation, which offers a
somewhat more explict path to validation functions.

Pydantic V2 validation

While **pydantic** version 2 will offer a `@field_validator` decorator that's very similar to
the legacy `@validator` decorator, this approach suffers from an irksome problem. It can
be confusing to have the decorator listing the fields to which the validation rule applies.
Some confusion can arise because of the separation between the field definition and the
function that validates the values for the field. In our example class, the validator applies
to the x and y fields, a detail that might be difficult to spot when first looking at the class.

The newer design pattern for the analysis model class is shown in the following example
for Pydantic version 2:

```
from pydantic import BaseModel
from pydantic.functional_validators import field_validator, BeforeValidator

from typing import Annotated
```

```
def clean_value(value: str | float) -> str | float:
    match value:
        case str():
            for char in "\N{ZERO WIDTH SPACE}":
                value = value.replace(char, "")
            return value
        case float():
            return value


class SeriesSample(BaseModel):
    x: Annotated[float, BeforeValidator(clean_value)]
    y: Annotated[float, BeforeValidator(clean_value)]
```

We've omitted the `from_acquire_dataclass()` method definition, since it doesn't change.

The cleaning function is defined outside the class, making it more easily reused in a complicated application where a number of rules may be widely reused in several models. The `Annotated[]` type hint combines the base type with a sequence of validator objects. In this example, the base type is `float` and the validator objects are `BeforeValidator` objects that contain the function to apply.

To reduce the obvious duplication, a `TypeAlias` can be used. For example,

```
from typing import Annotated, TypeAlias
```

```
CleanFloat: TypeAlias = Annotated[float, BeforeValidator(clean_value)]
```

Using an alias permits the model to use the type hint `CleanFloat`.
For example `x: CleanFloat`.

Further, the `Annotated` hints are composable. An annotation can add features to a previously-defined annotation. This ability to build more sophisticated annotations on top of foundational annotations offers a great deal of promise for defining classes in a succinct and expressive fashion.

Now that we've seen how to implement a single validation, we need to consider the alternatives, and how many different kinds of validation functions an application might need.

Validation function design

The pydantic package offers a vast number of built-in conversions based entirely on annotations. While these can cover a large number of common cases, there are still some situations that require special validators, and perhaps even special type definitions.

In *Conversions and processing*, we considered some of the kinds of processing that might be required. These included the following kinds of conversions:

- Decomposing source fields into their atomic components.

- Merging separated source fields to create proper value. This is common with dates and times, for example.

- Multiple subentities may be present in a feed of samples. This can be called a discriminated union: the feed as a whole is a unique of disjoint types, and a discriminator value (or values) distinguishes the various subtypes.

- A field may be a "token" used to deidentify something about the original source. For example, a replacement token for a driver's license number may replace the real government-issued number to make the individual anonymous.

Additionally, we may have observability considerations that lead us to write our own a unique validator that can write needed log entries or update counters showing how many times a particular validation found problems. This enhanced visibility can help pinpoint problems with data that is often irregular or suffers from poor quality control.

We'll dive into these concepts more deeply in *Chapter 10, Data Cleaning Features*. In *Chapter 10, Data Cleaning Features*, we'll also look at features for handling primary and foreign keys. For now, we'll focus on the built-in type conversion functions that are part of Python's built-in functions, and the standard library. But we need to recognize that there are going to be extensions and exceptions.

We'll look at the overall design approach in the next section.

Incremental design

The design of the cleaning application is difficult to finalize without detailed knowledge of the source data. This means the cleaning application depends on lessons learned by making a data inspection notebook. One idealized workflow begins with "understand the requirements" and proceeds to "write the code," treating these two activities as separate, isolated steps. This conceptual workflow is a bit of a fallacy. It's often difficult to understand the requirements without a detailed examination of the actual source data to reveal the quirks and oddities that are present. The examination of the data often leads to the first drafts of data validation functions. In this case, the requirements will take the form of draft versions of the code, not a carefully-crafted document.

This leads to a kind of back-and-forth between *ad-hoc* inspection and a formal data cleaning application. This iterative work often leads to a module of functions to handle the problem domain's data. This module can be shared by inspection notebooks as well as automated applications. Proper engineering follows the **DRY (Don't Repeat Yourself)** principle: code should not be copied and pasted between modules. It should be put into a shared module so it can be reused properly.

In some cases, two data cleaning functions will be similar. Finding this suggests some kind of decomposition is appropriate to separate the common parts from the unique parts. The redesign and refactoring are made easier by having a suite of unit tests to confirm that no old functionality was broken when the functions were transformed to remove duplicated code.

The work of creating cleaning applications is iterative and incremental. Rare special cases are — well — rare, and won't show up until well after the processing pipeline seems finished.

The unexpected arrival special case data is something like birders seeing a bird outside its expected habitat. It helps to think of a data inspection notebook like a bird watcher's immense spotting scope, used to look closely at one unexpected, rare bird, often in a flock of birds with similar feeding and roosting preferences. The presence of the rare bird becomes a new datapoint for ornithologists (and amateur enthusiasts). In the case of unexpected data, the inspection notebook's lessons become a new code for the conversions module.

The overall main module in the data cleaning application will implement the **command-line interface (CLI)**. We'll look at this in the next section.

CLI application

The UX for this application suggests that it operates in the following distinct contexts:

- As a standalone application. The user runs the `src/acquire.py` program. Then, the user runs the `src/clean.py` program.

- As a stage in a processing pipeline. The user runs a shell command that pipes the output from the `src/acquire.py` program into the `src/clean.py` program. This is the subject of *Project 3.6: Integration to create an acquisition pipeline*.

This leads to the following two runtime contexts:

- When the application is provided an input path, it's being used as a stand-alone application.

- When no input path is provided, the application reads from `sys.stdin`.

A similar analysis can apply to the **acquire** application. If an output path is provided, the application creates and writes the named file. If no output path is provided, the application writes to `sys.stdout`.

One essential consequence of this is all logging **must** be written to `sys.stderr`.

Use **stdin** and **stdout** exclusively for application data, nothing else.

Use a consistent, easy-to-parse text format like ND JSON for application data.

Use **stderr** as the destination for all control and error messages.

This means print() may require the file=sys.stderr to direct debugging output to **stderr**. Or, avoid simple print() and use logger.debug() instead.

For this project, the stand-alone option is all that's needed. However, it's important to understand the alternatives that will be added in later projects. See *Project 3.6: Integration to create an acquisition pipeline* for this more tightly-integrated alternative.

Redirecting stdout

Python provides a handy tool for managing the choice between "write to an open file" and "write to **stdout**". It involves the following essential design principle.

 Always provide file-like objects to functions and methods processing data.

This suggests a data-cleaning function like the following:

```
from typing import TextIO


def clean_all(acquire_file: TextIO, analysis_file: TextIO) -> None:
    . . .
```

This function can use json.loads() to parse each document from the acquire_file. It uses json.dumps() to save each document to the analysis_file to be used for later analytics.

The overall application can then make a choice among four possible ways to use this clean_all() function:

- **Stand-alone**: This means `with` statements manage the open files created from the `Path` names provided as argument values.

- **Head of a pipeline**: A `with` statement can manage an open file passed to `acquire_file`. The value of `analysis_file` is `sys.stdout`.

- **Tail of a pipeline**: The acquired input file is `sys.stdin`. A `with` statement manages an open file (in write mode) for the `analysis_file`.

- **Middle of a pipeline**: The `acquire_file` is `sys.stdin`; the `analysis_file` is `sys.stdout`.

Now that we've looked at a number of technical approaches, we'll turn to the list of deliverables for this project in the next section.

Deliverables

This project has the following deliverables:

- Documentation in the `docs` folder.

- Acceptance tests in the `tests/features` and `tests/steps` folders.

- Unit tests for the application modules in the `tests` folder.

- Application to clean some acquired data and apply simple conversions to a few fields. Later projects will add more complex validation rules.

We'll look at a few of these deliverables in a little more detail.

When starting a new kind of application, it often makes sense to start with acceptance tests. Later, when adding features, the new acceptance tests may be less important than new unit tests for the features. We'll start by looking at a new scenario for this new application.

Acceptance tests

As we noted in *Chapter 4, Data Acquisition Features: Web APIs and Scraping*, we can provide a large block of text as part of a Gherkin scenario. This can be the contents of an input file. We can consider something like the following scenario.

```
Scenario: Valid file is recognized.
    Given a file "example_1.ndjson" with the following content
        """
        {"x": "1.2", "y": "3.4"}
        {"x": "five", "z": null}
        """
    When the clean tool is run
    Then the output shows 1 good records
    And the output shows 1 faulty records
```

This kind of scenario lets us define source documents with valid data. We can also define source documents with invalid data.

We can use the Then steps to confirm additional details of the output. For example, if we've decided to make all of the cleaning operations visible, the test scenario can confirm the output contains all of the cleanup operations that were applied.

The variety of bad data examples and the number of combinations of good and bad data suggest there can be a lot of scenarios for this kind of application. Each time new data shows up that is acquired, but cannot be cleaned, new examples will be added to these acceptance test cases.

It can, in some cases, be very helpful to publish the scenarios widely so all of the stakeholders can understand the data cleaning operations. The Gherkin language is designed to make it possible for people with limited technical skills to contribute to the test cases.

We also need scenarios to run the application from the command-line. The When step definition for these scenarios will be subprocess.run() to invoke the **clean** application, or to invoke a shell command that includes the **clean** application.

Unit tests for the model features

It's important to have automated unit tests for the model definition classes.

It's also important to **not** test the pydantic components. We don't, for example, need to test the ordinary string-to-float conversions the `pydantic` module already does; we can trust this works perfectly.

We **must** test the validator functions we've written. This means providing test cases to exercise the various features of the validators. Additionally, any overall `from_acquire_dataclass()` method needs to have test cases.

Each of these test scenarios works with a given acquired document with the raw data. When the `from_acquire_dataclass()` method is evaluated, then there may be an exception or a resulting analytic model document is created.

The exception testing can make use of the `pytest.raises()` context manager. The test is written using a `with` statement to capture the exception.

See `https://docs.pytest.org/en/7.2.x/how-to/assert.html#assertions-about-exp` `ected-exceptions` for examples.

Of course, we also need to test the processing that's being done. By design, there isn't very much processing involved in this kind of application. The bulk of the processing can be only a few lines of code to consume the raw model objects and produce the analytical objects. Most of the work will be delegated to modules like `json` and `pydantic`.

Application to clean data and create an NDJSON interim file

Now that we have acceptance and unit test suites, we'll need to create the `clean` application. Initially, we can create a place-holder application, just to see the test suite fail. Then we can fill in the various pieces until the application – as a whole – works.

Flexibility is paramount in this application. In the next chapter, *Chapter 10, Data Cleaning Features*, we will introduce a large number of data validation scenarios. In *Chapter 11, Project 3.7: Interim Data Persistence* we'll revisit the idea of saving the cleaned data. For now, it's imperative to create clean data; later, we can consider what format might be best.

Summary

This chapter has covered a number of aspects of data validation and cleaning applications:

- CLI architecture and how to design a simple pipeline of processes.

- The core concepts of validating, cleaning, converting, and standardizing raw data.

In the next chapter, we'll dive more deeply into a number of data cleaning and standardizing features. Those projects will all build on this base application framework. After those projects, the next two chapters will look a little more closely at the analytical data persistence choices, and provide an integrated web service for providing cleaned data to other stakeholders.

Extras

Here are some ideas for you to add to this project.

Create an output file with rejected samples

In *Error reports* we suggested there are times when it's appropriate to create a file of rejected samples. For the examples in this book — many of which are drawn from well-curated, carefully managed data sets — it can feel a bit odd to design an application that will reject data.

For enterprise applications, data rejection is a common need.

It can help to look at a data set like this: `https://datahub.io/core/co2-ppm`. This contains data same with measurements of CO_2 levels measures with units of ppm, parts per million.

This has some samples with an invalid number of days in the month. It has some samples where a monthly CO_2 level wasn't recorded.

It can be insightful to use a rejection file to divide this data set into clearly usable records, and records that are not as clearly usable.

The output will **not** reflect the analysis model. These objects will reflect the acquire model;

they are the items that would not convert properly from the acquired structure to the desired analysis structure.

10

Data Cleaning Features

There are a number of techniques for validating and converting data to native Python objects for subsequent analysis. This chapter guides you through three of these techniques, each appropriate for different kinds of data. The chapter moves on to the idea of standardization to transform unusual or atypical values into a more useful form. The chapter concludes with the integration of acquisition and cleansing into a composite pipeline.

This chapter will expand on the project in *Chapter 9, Project 3.1: Data Cleaning Base Application*. The following additional skills will be emphasized:

- CLI application extension and refactoring to add features.

- Pythonic approaches to validation and conversion.

- Techniques for uncovering key relationships.

- Pipeline architectures. This can be seen as a first step toward a processing **DAG** (**Directed Acyclic Graph**) in which various stages are connected.

We'll start with a description of the first project to expand on the previous chapters on processing. This will include some new **Pydantic** features to work with more complex data source fields.

Project 3.2: Validate and convert source fields

In *Chapter 9, Project 3.1: Data Cleaning Base Application* we relied on the foundational behavior of the **Pydantic** package to convert numeric fields from the source text to Python types like int, float, and Decimal. In this chapter, we'll use a dataset that includes date strings so we can explore some more complex conversion rules.

This will follow the design pattern from the earlier project. It will use a distinct data set, however, and some unique data model definitions.

Description

This project's intent is to perform data validation, cleaning, and standardization. This project will expand on the features of the pydantic package to do somewhat more sophisticated data validations and conversions.

This new data cleaning application can be designed around a data set like https://tide sandcurrents.noaa.gov/tide_predictions.html. The tide predictions around the US include dates, but the fields are decomposed, and our data cleaning application needs to combine them.

For a specific example, see https://tidesandcurrents.noaa.gov/noaatideannual.htm l?id=8725769. Note that the downloaded .txt file is a tab-delimited CSV file with a very complicated multi-line header. This will require some sophisticated acquisition processing similar to the examples shown in *Chapter 3, Project 1.1: Data Acquisition Base Application*.

An alternative example is the CO_2 PPM — Trends in Atmospheric Carbon Dioxide data set, available at https://datahub.io/core/co2-ppm. This has dates that are provided in two forms: as a year-month-day string and as a decimal number. We can better understand this data if we can reproduce the decimal number value.

The second example data set is `https://datahub.io/core/co2-ppm/r/0.html` This is an HTML file, requiring some acquisition processing similar to the examples from *Chapter 4, Data Acquisition Features: Web APIs and Scraping.*

The use case for this cleaning application is identical to the description shown in *Chapter 9, Project 3.1: Data Cleaning Base Application.* The acquired data — pure text, extracted from the source files — will be cleaned to create **Pydantic** models with fields of useful Python internal types.

We'll take a quick look at the tide table data on the `https://tidesandcurrents.noaa.gov` website.

```
NOAA/NOS/CO-OPS
Disclaimer: These data are based upon the latest information available ...
Annual Tide Predictions
StationName: EL JOBEAN, MYAKKA RIVER
State: FL
Stationid: 8725769
ReferencedToStationName: St. Petersburg, Tampa Bay
ReferencedToStationId: 8726520
HeightOffsetLow: * 0.83
HeightOffsetHigh: * 0.83
TimeOffsetLow: 116
TimeOffsetHigh: 98
Prediction Type: Subordinate
From: 20230101 06:35 - 20231231 19:47
Units: Feet and Centimeters
Time Zone: LST_LDT
Datum: MLLW
Interval Type: High/Low


Date  Day Time Pred(Ft) Pred(cm) High/Low
```

```
2023/01/01 Sun 06:35 AM -0.13 -4 L
2023/01/01 Sun 01:17 PM 0.87 27 H
etc.
```

The data to be acquired has two interesting structural problems:

1. There's a 19-line preamble containing some useful metadata. Lines 2 to 18 have a format of a label and a value, for example, `State: FL`.

2. The data is tab-delimited CSV data. There appear to be six column titles. However, looking at the tab characters, there are eight columns of header data followed by nine columns of data.

The acquired data should fit the dataclass definition shown in the following fragment of a class definition:

```python
from dataclasses import dataclass


@dataclass
class TidePrediction:
    date: str
    day: str
    time: str
    pred_ft: str
    pred_cm: str
    high_low: str


    @classmethod
    def from_row(
        cls: type["TidePrediction"],
        row: list[str]
    ) -> "TidePrediction":

        ...
```

The example omits the details of the `from_row()` method. If a CSV reader is used, this method needs to pick out columns from the CSV-format file, skipping over the generally empty columns. If regular expressions are used to parse the source lines, this method will use the groups from the match object.

Since this looks like many previous projects, we'll look at the distinct technical approach next.

Approach

The core processing of the data cleaning application should be — except for a few module changes — very similar to the earlier examples. For reference, see *Chapter 9, Project 3.1: Data Cleaning Base Application*, specifically *Approach*. This suggests that the clean module should have minimal changes from the earlier version.

The principle differences should be two different implementations of the `acquire_model` and the `analysis_model`. For the tide data example, a class is shown in the *Description* section that can be used for the acquire model.

> It's important to maintain a clear distinction between the acquired data, which is often text, and the data that will be used for later analysis, which can be a mixture of more useful Python object types.

> The two-step conversion from source data to the interim acquired data format, and from the acquired data format to the clean data format can — sometimes — be optimized to a single conversion.

> An optimization to combine processing into a single step can also make debugging more difficult.

We'll show one approach to defining the enumerated set of values for the state of the tide. In the source data, codes of `'H'` and `'L'` are used. The following class will define this enumeration of values:

```
from enum import StrEnum
```

```python
class HighLow(StrEnum):
    high = 'H'
    low = 'L'
```

We'll rely on the enumerated type and two other annotated types to define a complete record. We'll return to the annotated types after showing the record as a whole first. A complete tide prediction record looks as follows:

```python
import datetime
from typing import Annotated, TypeAlias


from pydantic import BaseModel
from pydantic.functional_validators import AfterValidator, BeforeValidator


# See below for the type aliases.


class TidePrediction(BaseModel):
    date: TideCleanDateTime
    pred_ft: float
    pred_cm: float
    high_low: TideCleanHighLow


    @classmethod
    def from_acquire_dataclass(
            cls,
            acquired: acquire_model.TidePrediction
    ) -> "TidePrediction":
        source_timestamp = f"{acquired.date} {acquired.time}"
        return TidePrediction(
            date=source_timestamp,
```

```
            pred_ft=acquired.pred_ft,
            pred_cm=acquired.pred_cm,
            high_low=acquired.high_low
    )
```

This shows how the source columns' date and time are combined into a single text value prior to validation. This is done by the from_acquire_dataclass() method, so it happens before invoking the TidePrediction constructor.

The TideCleanDateTime and TideCleanHighLow type hints will leverage annotated types to define validation rules for each of these attributes. Here are the two definitions:

```
TideCleanDateTime: TypeAlias = Annotated[
    datetime.datetime, BeforeValidator(clean_date)]
TideCleanHighLow: TypeAlias = Annotated[
    HighLow, BeforeValidator(lambda text: text.upper())]
```

The TideCleanDateTime type uses the clean_date() function to clean up the date string prior to any attempt at conversion. Similarly, the TideCleanHighLow type uses a lambda to transform the value to upper case before validation against the HighLow enumerated type.

The clean_date() function works by applying the one (and only) expected date format to the string value. This is not designed to be flexible or permissive. It's designed to confirm the data is an exact match against expectations.

The function looks like this:

```
def clean_date(v: str | datetime.datetime) -> datetime.datetime:
    match v:
        case str():
            return datetime.datetime.strptime(v, "%Y/%m/%d %I:%M %p")
        case _:
            return v
```

If the data doesn't match the expected format, the strptime() function will raise a

`ValueError` exception. This will be incorporated into a `pydantic.ValidationError` exception that enumerates all of the errors encountered. The `match` statement will pass non-string values through to the **pydantic** handler for validation; we don't need to handle any other types.

This model can also be used for analysis of clean data. (See _Chapter 13, Project 4.1: Visual Analysis Techniques._) In this case, the data will already be a valid `datetime.datetime` object, and no conversion will need to be performed. The use of a type hint of `str | datetime.datetime` emphasizes the two types of values this method will be applied to.

This two-part "combine and convert" operation is broken into two steps to fit into the **Pydantic** design pattern. The separation follows the principle of minimizing complex initialization processing and creating class definitions that are more declarative and less active.

 It's often helpful to keep the conversion steps small and separate.

Premature optimization to create a single, composite function is often a nightmare when changes are required.

For display purposes, the date, day-of-week, and time-of-day can be extracted from a single `datetime` instance. There's no need to keep many date-related fields around as part of the `TidePrediction` object.

The tide prediction is provided in two separate units of measure. For the purposes of this example, we retained the two separate values. Pragmatically, the height in feet is the height in cm multiplied by $\frac{1}{30.48}$.

For some applications, where the value for height in feet is rarely used, a property might make more sense than a computed value. For other applications, where the two heights are both used widely, having both values computed may improve performance.

Deliverables

This project has the following deliverables:

- Documentation in the `docs` folder.

- Acceptance tests in the `tests/features` and `tests/steps` folders.

- Unit tests for the application modules in the `tests` folder.

- Application to clean some acquired data and apply simple conversions to a few fields. Later projects will add more complex validation rules.

Many of these deliverables are described in previous chapters. Specifically, *Chapter 9, Project 3.1: Data Cleaning Base Application* covers the basics of the deliverables for this project.

Unit tests for validation functions

The unique validators used by a Pydantic class need test cases. For the example shown, the validator function is used to convert two strings into a date.

Boundary Value Analysis suggests there are three equivalence classes for date conversions:

- Syntactically invalid dates. The punctuation or the number of digits is wrong.

- Syntactically valid, but calendrical invalid dates. The 30th of February, for example, is invalid, even when formatted properly.

- Valid dates.

The above list of classes leads to a minimum of three test cases.

Some developers like to explore each of the fields within a date, providing 5 distinct values: the lower limit (usually 1), the upper limit (e.g., 12 or 31), just below the limit (e.g., 0), just above the upper limit (e.g., 13 or 32), and a value that's in the range and otherwise valid. These additional test cases, however, are really testing the `strptime()` method of the `datetime` class. These cases are duplicate tests of the `datetime` module. These cases are *not* needed, since the `datetime` module already has plenty of test cases for calendrically invalid date strings.

 Don't test the behavior of modules outside the application. Those modules have their own test cases.

In the next section, we'll look at a project to validate nominal data. This can be more complicated than validating ordinal or cardinal data.

Project 3.3: Validate text fields (and numeric coded fields)

For nominal data, we'll use **pydantic**'s technique of applying a validator function to the value of a field. In cases where the field contains a code consisting only of digits, there can be some ambiguity as to whether or not the value is a cardinal number. Some software may treat any sequence of digits as a number, dropping leading zeroes. This can lead to a need to use a validator to recover a sensible value for fields that are strings of digits, but not cardinal values.

Description

This project's intent is to perform data validation, cleaning, and standardization. This project will expand on the features of the **Pydantic** package to do somewhat more sophisticated data validation and conversion.

We'll continue working with a data set like `https://tidesandcurrents.noaa.gov/ti de_predictions.html`. The tide predictions around the US include dates, but the date is decomposed into three fields, and our data cleaning application needs to combine them.

For a specific example, see `https://tidesandcurrents.noaa.gov/noaatideannual.htm l?id=8725769`. Note that the downloaded `.txt` file is really a tab-delimited CSV file with a complex header. This will require some sophisticated acquisition processing similar to the examples shown in *Chapter 3, Project 1.1: Data Acquisition Base Application*.

For data with a relatively small domain of unique values, a Python `enum` class is a very handy way to define the allowed collection of values. Using an enumeration permits simple, strict validation by `pydantic`.

Some data — like account numbers, as one example — have a large domain of values that may be in a state of flux. Using an `enum` class would mean transforming the valid set of account numbers into an enumerated type before attempting to work with any data. This may not be particularly helpful, since there's rarely a compelling need to confirm that an account number is valid; this is often a stipulation that is made about the data.

For fields like account numbers, there can be a need to validate potential values without an enumeration of all allowed values. This means the application must rely on patterns of the text to determine if the value is valid, or if the value needs to be cleaned to make it valid. For example, there may be a required number of digits, or check digits embedded within the code. In the case of credit card numbers, the last digit of a credit card number is used as part of confirmation that the overall number is valid. For more information, see `https://www.creditcardvalidator.org/articles/luhn-algorithm`.

After considering some of the additional validations that need to be performed, we'll take a look at a design approach for adding more complicated validations to the cleaning application.

Approach

For reference to the general approach to this application, see *Chapter 9, Project 3.1: Data Cleaning Base Application*, specifically *Approach*.

The model can be defined using the **pydantic** package. This package offers two paths to validating string values against a domain of valid values. These alternatives are:

- Define an enumeration with all valid values.

- Define a regular expression for the string field. This has the advantage of defining very large domains of valid values, including *potentially* infinite domains of values.

Enumeration is an elegant solution that defines the list of values as a class. As shown earlier, it might look like this:

```
import enum


class HighLow(StrEnum):
    high = 'H'
    low = 'L'
```

This will define a domain of two string values, "L" and "H". This map provides easier-to-understand names, Low and High. This class can be used by **pydantic** to validate a string value.

An example of a case when we need to apply a BeforeValidator annotated type might be some tide data with lower-case "h" and "l" instead of proper upper-case "H" or "L". This allows the validator to clean the data **prior** to the built-in data conversion.

We might use an annotated type. It looked like this in the preceding example:

```
TideCleanHighLow: TypeAlias = Annotated[
    HighLow, BeforeValidator(lambda text: text.upper())]
```

The annotated type hint describes the base type, HighLow, and a validation rule to be applied before the **pydantic** conversion. In this case, it's a lambda to convert the text to upper case. We've emphasized the validation of enumerated values using an explicit enumeration because it is an important technique for establishing the complete set of allowed codes for a given attribute. The enumerated type's class definition is often a handy place to record notes and other information about the coded values.

Now that we've looked at the various aspects of the approach, we can turn our attention to the deliverables for this project.

Deliverables

This project has the following deliverables:

- Documentation in the docs folder.

- Acceptance tests in the tests/features and tests/steps folders.

- Unit tests for the application modules in the `tests` folder.

- Application to clean source data in a number of fields.

Many of these deliverables are described in previous chapters. Specifically, *Chapter 9, Project 3.1: Data Cleaning Base Application* covers the basics of the deliverables for this project.

Unit tests for validation functions

The unique validators used by a pydantic class need test cases. For the example shown, the validator function is used to validate the state of the tide. This is a small domain of enumerated values. There are three core kinds of test cases:

- Valid codes like `'H'` or `'L'`.

- Codes that can be reliably cleaned. For example, lower-case codes `'h'` and `'l'` are unambiguous. A data inspection notebook may reveal non-code values like `'High'` or `'Low'`, also. These can be reliably cleaned.

- Invalid codes like `' '`, or `'9'`.

The domain of values that can be cleaned properly is something that is subject to a great deal of change. It's common to find problems and use an inspection notebook to uncover a new encoding when upstream applications change. This will lead to additional test cases, and then additional validation processing to make the test cases pass.

In the next project, we'll look at the situation where data must be validated against an externally defined set of values.

Project 3.4: Validate references among separate data sources

In *Chapter 9, Project 3.1: Data Cleaning Base Application* we relied on the foundational behavior of Pydantic to convert fields from source text to Python types. This next project will look at a more complicated validation rule.

Description

This project's intent is to perform data validation, cleaning, and standardization. This project will expand on the features of the `pydantic` package to do somewhat more sophisticated data validations and conversions.

Data sets in `https://data.census.gov` have **ZIP Code Tabulation Areas (ZCTAs)**. For certain regions, these US postal codes can (and should) have leading zeroes. In some variations on this data, however, the ZIP codes get treated as numbers and the leading zeroes get lost.

Data sets at `https://data.census.gov` have information about the city of Boston, Massachusets, which has numerous US postal codes with leading zeroes. The Food Establishment Inspections available at `https://data.boston.gov/group/permitting` provides insight into Boston-area restaurants. In addition to postal codes (which are nominal data), this data involves numerous fields that contain nominal data as well as ordinal data.

For data with a relatively small domain of unique values, a Python `enum` class is a very handy way to define the allowed collection of values. Using an enumeration permits simple, strict validation by **Pydantic**.

Some data — like account numbers, as one example — have a large domain of values that may be in a state of flux. Using an `enum` class would mean transforming the valid set of account numbers into an enum before attempting to work with any data. This may not be particularly helpful, since there's rarely a compelling need to confirm that an account number is valid; this is often a simple stipulation that is made about the data.

This leads to a need to validate potential values without an enumeration of the allowed values. This means the application must rely on patterns of the text to determine if the value is valid, or if the value needs to be cleaned to make it valid.

When an application cleans postal code data, there are two distinct parts to the cleaning:

1. Clean the postal code to have the proper format. For US ZIP codes, it's generally 5

digits. Some codes are 5 digits, a hyphen, and 4 more digits.

2. Compare the code with some master list to be sure it's a meaningful code that references an actual post office or location.

It's important to keep these separate since the first step is covered by the previous project, and doesn't involve anything terribly complicated. The second step involves some additional processing to compare a given record against a master list of allowed values.

Approach

For reference to the general approach to this application, see *Chapter 9, Project 3.1: Data Cleaning Base Application*, specifically *Approach*.

When we have nominal values that must refer to external data, we can call these "foreign keys." They're references to an external collection of entities for which the values are primary keys. An example of this is a postal code. There is a defined list of valid postal codes; the code is a primary key in this collection. In our sample data, the postal code is a foreign key reference to the defining collection of postal codes.

Other examples include country codes, US state codes, and US phone system area codes. We can write a regular expression to describe the potential domain of key values. For US state codes, for example, we can use the regular expression r'\w\w' to describe state codes as having two letters. We could narrow this domain slightly using r'[A-Z]{2}' to require the state code use upper-case letters only. There are only 50 state codes, plus a few territories and districts; limiting this further would make for a very long regular expression.

The confounding factor here is when the primary keys need to be loaded from an external source — for example, a database. In this case, the simple @validator method has a dependency on external data. Further, this data must be loaded prior to any data cleaning activities.

We have two choices for gathering the set of valid key values:

- Create an Enum class with a list of valid values.

- Define a `@classmethod` to initialize the pydantic class with valid values.

For example, `https://data.opendatasoft.com` has a useful list of US zip codes. See the URL `https://data.opendatasoft.com/api/explore/v2.1/catalog/datasets/georef-united-states-of-america-zc-point@public/exports/csv` for US Zip Codes Points, United States of America. This is a file that can be downloaded and transformed into an enum or used to initialize a class. The `Enum` class creation is a matter of creating a list of two tuples with the label and the value for the enumeration. The `Enum` definition can be built with code like the following example:

```python
import csv
import enum
from pathlib import Path


def zip_code_values() -> list[tuple[str, str]]:
    source_path = Path.home() / "Downloads" / "georef-united-states-of-
        america-zc-point@public.csv"
    with source_path.open(encoding='utf_8_sig') as source:
        reader = csv.DictReader(source, delimiter=';')
        values = [
            (f"ZIP_{zip['Zip Code']:0>5s}", f"{zip['Zip Code']:0>5s}")
            for zip in reader
        ]
    return values


ZipCode = enum.Enum("ZipCode", zip_code_values())
```

This will create an enumerated class, `ZipCode`, from the approximately 33,000 ZIP codes in the downloaded source file. The enumerated labels will be Python attribute names similar to `ZIP_75846`. The values for these labels will be the US postal codes, for example, `'75846'`. The `":0>5s"` string format will force in leading zeroes where needed.

The `zip_code_values()` function saves us from writing 30,000 lines of code to define the enumeration class, `ZipCode`. Instead, this function reads 30,000 values, creating a list of pairs used to create an `Enum` subclass.

The odd encoding of `utf_8_sig` is necessary because the source file has a leading **byte-order mark (BOM)**. This is unusual butpermitted by Unicode standards. Other data sources for ZIP codes may not include this odd artifact. The encoding gracefully ignores the BOM bytes.

> The unusual encoding of `utf_8_sig` is a special case because this file happens to be in an odd format.
>
> There are a large number of encodings for text. While UTF-8 is popular, it is not universal.
>
> When unusual characters appear, it's important to find the source of the data and ask what encoding they used.
>
> In general, it's impossible to uncover the encoding given a sample file. There are a large number of valid byte code mappings that overlap between ASCII, CP1252, and UTF-8.

This design requires the associated data file. One potential improvement is to create a Python module from the source data.

Using the **Pydantic** functional validators uses a similar algorithm to the one shown above. The validation initialization is used to build an object that retains a set of valid values. We'll start with the goal of a small model using annotated types. The model looks like this:

```python
import csv
from pathlib import Path
import re
from typing import TextIO, TypeAlias, Annotated
```

```python
from pydantic import BaseModel, Field
from pydantic.functional_validators import BeforeValidator, AfterValidator


# See below for the type aliases.


ValidZip: TypeAlias = Annotated[
    str,
    BeforeValidator(zip_format_valid),
    AfterValidator(zip_lookup_valid)]


class SomethingWithZip(BaseModel):
    # Some other fields
    zip: ValidZip
```

The model relies on the `ValidZip` type. This type has two validation rules: before any conversion, a `zip_format_valid()` function is applied, and after conversion, a `zip_lookup_valid()` function is used.

We've only defined a single field, `zip`, in this **Pydantic** class. This will let us focus on the validation-by-lookup design. A more robust example, perhaps based on the Boston health inspections shown above, would have a number of additional fields reflecting the source data to be analyzed.

The before validator function, `zip_format_valid()`, compares the ZIP code to a regular expression to ensure that it is valid:

```python
def zip_format_valid(zip: str) -> str:
    assert re.match(r'\d{5}|\d{5}-d{4}', zip) is not None,
    f"ZIP invalid format {zip!r}"
    return zip
```

The `zip_format_valid()` can be expanded to use an f-string like `f"{zip:0>5s}` to reformat a ZIP code that's missing the leading zeroes. We'll leave this for you to integrate into this function.

The after validator function is a callable object. It's an instance of a class that defines the __call__() method.

Here's the core class definition, and the creation of the instance:

```python
class ZipLookupValidator:
    """Compare a code against a list."""
    def __init__(self) -> None:
        self.zip_set: set[str] = set()

    def load(self, source: TextIO) -> None:
        reader = csv.DictReader(source, delimiter=';')
        self.zip_set = {
            f"{zip['Zip Code']:0>5s}"
            for zip in reader
        }

    def __call__(self, zip: str) -> str:
        if zip in self.zip_set:
            return zip
        raise ValueError(f"invalid ZIP code {zip}")


zip_lookup_valid = ZipLookupValidator()
```

This will define the zip_lookup_valid callable object. Initially, there's new value for the internal self.zip_set attribute. This must be built using a function that evaluates zip_lookup_valid.load(source). This will populate the set of valid values.

We've called this function prepare_validator() and it looks like this:

```python
def prepare_validator() -> None:
    source_path = (
```

```
        Path.home() / "Downloads" /
        "georef-united-states-of-america-zc-point@public.csv"
    )
    with source_path.open(encoding='utf_8_sig') as source:
        zip_lookup_valid.load(source)
```

This idea of a complex validation follows the SOLID design principle. It separates the essential work of the SomethingWithZip class from the ValidZip type definition.

Further, the ValidZip type depends on a separate class, ZipLookupValidator, which handles the complications of loading data. This separation makes it somewhat easier to change validation files, or change the format of the data used for validation without breaking the SomethingWithZip class and the applications that use it. Further, it provides a reusable type, ValidZip. This can be used for multiple fields of a model, or multiple models.

Having looked at the technical approach, we'll shift to looking at the deliverables for this project.

Deliverables

This project has the following deliverables:

- Documentation in the docs folder.

- Acceptance tests in the tests/features and tests/steps folders.

- Unit tests for the application modules in the tests folder.

- Application to clean and validate data against external sources.

Many of these deliverables are described in previous chapters. Specifically, *Chapter 9, Project 3.1: Data Cleaning Base Application* covers the basics of the deliverables for this project.

Unit tests for data gathering and validation

The unique validators used by a **Pydantic** class need test cases. For the example shown, the validator function is used to validate US ZIP codes. There are three core kinds of test cases:

- Valid ZIP codes with five digits that are found in the ZIP code database.

- Syntactically valid ZIP codes with five digits that are **not** found in the ZIP code database.

- Syntactically invalid ZIP that don't have five digits, or can't — with the addition of leading zeroes — be made into valid codes.

Project 3.5: Standardize data to common codes and ranges

Another aspect of cleaning data is the transformation of raw data values into standardized values. For example, codes in use have evolved over time, and older data codes should be standardized to match new data codes. The notion of standardizing values can be a sensitive topic if critical information is treated as an outlier and rejected or improperly standardized.

We can also consider imputing new values to fill in for missing values as a kind of standardization technique. This can be a necessary step when dealing with missing data or data that's likely to represent some measurement error, not the underlying phenomenon being analyzed.

This kind of transformation often requires careful, thoughtful justification. We'll show some programming examples. The deeper questions of handling missing data, imputing values, handling outliers, and other standardization operations are outside the scope of this book.

See `https://towardsdatascience.com/6-different-ways-to-compensate-for-missing-values-data-imputation-with-examples-6022d9ca0779` for an overview of some ways to deal with missing or invalid data.

Description

Creating standardized values is at the edge of data cleaning and validation. These values can be described as "derived" values, computed from existing values.

There are numerous kinds of standardizations; we'll look at two:

1. Compute a standardized value, or Z-score, for cardinal data. For a normal distribution, the Z-scores have a mean of 0, and a standard deviation of 1. It permits comparing values measured on different scales.

2. Collapse nominal values into a single standardized value. For example, replacing a number of historical product codes with a single, current product code.

The first of these, computing a Z-score, rarely raises questions about the statistical validity of the standardized value. The computation, $Z = \frac{x-\mu}{\sigma}$, is well understood and has known statistical properties.

The second standardization, replacing nominal values with a standardized code, can be troubling. This kind of substitution may simply correct errors in the historical record. It may also obscure an important relationship. It's not unusual for a data inspection notebook to reveal outliers or erroneous values in a data set that needs to be standardized.

Enterprise software may have unrepaired bugs. Some business records can have unusual code values that map to other code values.

Of course, the codes in use may shift over time.

Some records may have values that reflect two eras: pre-repair and post-repair. Worse, of course, there may have been several attempts at a repair, leading to more nuanced timelines.

For this project, we need some relatively simple data. The Ancombe's Quartet data will do nicely as examples from which derived Z-scores can be computed. For more information, see *Chapter 3, Project 1.1: Data Acquisition Base Application.*

The objective is to compute standardized values for the two values that comprise the samples in the Anscombe's Quartet series. When the data has a normal distribution, these derived, standardized Z-scores will have a mean of zero and a standard deviation of one. When the data does not have a normal distribution, these values will diverge from the expected values.

Approach

For reference to the general approach to this application, see *Chapter 9, Project 3.1: Data Cleaning Base Application*, specifically *Approach.*

To replace values with preferred standardized values, we've seen how to clean bad data in previous projects. See, for example, *Project 3.3: Validate text fields (and numeric coded fields).*

For Z-score standardization, we'll be computing a derived value. This requires knowing the mean, μ, and standard deviation, σ, for a variable from which the Z-score can be computed.

This computation of a derived value suggests there are the following two variations on the analytical data model class definitions:

- An "initial" version, which lacks the Z-score values. These objects are incomplete and require further computation.

- A "final" version, where the Z-score values have been computed. These objects are complete.

There are two common approaches to handling this distinction between incomplete and complete objects:

- The two classes are distinct. The complete version is a subclass of the incomplete version, with additional fields defined.

- The derived values are marked as optional. The incomplete version starts with None values.

The first design is a more conventional object-oriented approach. The formality of a distinct type to clearly mark the state of the data is a significant advantage. The extra class definition, however, can be seen as clutter, since the incomplete version is transient data that doesn't create enduring value. The incomplete records live long enough to compute the complete version, and the file can then be deleted.

The second design is sometimes used for functional programming. It saves the subclass definition, which can be seen as a slight simplification.

```python
from pydantic import BaseModel


class InitialSample(BaseModel):
    x: float
    y: float


class SeriesSample(InitialSample):
    z_x: float
    z_y: float


    @classmethod
    def build_sample(cls, m_x: float, s_x: float,
    m_y: float, s_y: float, init:
      InitialSample)-> "SeriesSample":
        return SeriesSample(
            x=init.x, y=init.y,
            z_x=(init.x - m_x) / s_x,
            z_y=(init.y - m_y) / s_y
        )
```

These two class definitions show one way to formalize the distinction between the initially cleaned, validated, and converted data, and the complete sample with the standardized Z-scores present for both of the variables.

This can be handled as three separate operations:

1. Clean and convert the initial data, writing a temporary file of the `InitialSample` instances.

2. Read the temporary file, computing the means and standard deviations of the variables.

3. Read the temporary file again, building the final samples from the `InitialSample` instances and the computed intermediate values.

A sensible optimization is to combine the first two steps: clean and convert the data, accumulating values that can be used to compute the mean and standard deviation. This is helpful because the `statistics` module expects a sequence of objects that might not fit in memory. The mean, which involves a sum and a count, is relatively simple. The standard deviation requires accumulating a sum and a sum of squares.

$$m_x = \frac{\sum x}{n}$$

The mean of x, m_x, is the sum of the x values divided by the count of x values, shown as n.

$$s_x = \sqrt{\frac{\sum x^2 - \frac{(\sum x)^2}{n}}{n-1}}$$

The standard deviation of x, s_x, uses the sum of x^2, the sum of x, and the number of values, n.

This formula for the standard deviation has some numeric stability issues, and there are variations that are better designs. See `https://en.wikipedia.org/wiki/Algorithms_for_calculating_variance`.

We'll define a class that accumulates the values for mean and variance. From this, we can compute the standard deviation.

```
import math
```

```python
class Variance:
    def __init__(self):
        self.k: float | None = None
        self.e_x = 0.0
        self.e_x2 = 0.0
        self.n = 0

    def add(self, x: float) -> None:
        if self.k is None:
            self.k = x
        self.n += 1
        self.e_x += x - self.k
        self.e_x2 += (x - self.k) ** 2

    @property
    def mean(self) -> float:
        return self.k + self.e_x / self.n

    @property
    def variance(self) -> float:
        return (self.e_x2 - self.e_x ** 2 / self.n) / (self.n - 1)

    @property
    def stdev(self) -> float:
        return math.sqrt(self.variance)
```

This `variance` class performs an incremental computation of mean, standard deviation, and variance. Each individual value is presented by the `add()` method. After all of the data has been presented, the properties can be used to return the summary statistics.

It's used as shown in the following snippet:

```
var_compute = Variance()
for d in data:
    var_compute.add(d)


print(f"Mean = {var_compute.mean}")
print(f"Standard Deviation = {var_compute.stdev}")
```

This provides a way to compute the summary statistics without using a lot of memory. It permits the optimization of computing the statistics the first time the data is seen. And, it reflects a well-designed algorithm that is numerically stable.

Now that we've explored the technical approach, it's time to look at the deliverables that must be created for this project.

Deliverables

This project has the following deliverables:

- Documentation in the docs folder.

- Acceptance tests in the tests/features and tests/steps folders.

- Unit tests for the application modules in the tests folder.

- Application to clean the acquired data and compute derived standardized Z-scores.

Many of these deliverables are described in previous chapters. Specifically, *Chapter 9, Project 3.1: Data Cleaning Base Application* covers the basics of the deliverables for this project.

Unit tests for standardizing functions

There are two parts of the standardizing process that require unit tests. The first is the incremental computation of mean, variance, and standard deviation. This must be compared against results computed by the statistics module to assure that the results are correct.

The `pytest.approx` object (or the `math.isclose()` function) are useful for asserting the incremental computation matches the expected values from the standard library module.

Additionally, of course, the construction of the final sample, including the standardized Z-scores, needs to be tested. The test case is generally quite simple: a single value with a given x, y, mean of x, mean of y, the standard deviation of x, and the standard deviation of y need to be converted from the incomplete form to the complete form. The computation of the derived values is simple enough that the expected results can be computed by hand to check the results.

It's important to test this class, even though it seems very simple. Experience suggests that these seemingly simple classes are places where a + replaces a - and the distinction isn't noticed by people inspecting the code. This kind of small mistake is best found with a unit test.

Acceptance test

The acceptance test suite for this standardization processing will involve a main program that creates two output files. This suggests the after-scenario cleanup needs to ensure the intermediate file is properly removed by the application.

The cleaning application could use the `tempfile` module to create a file that will be deleted when closed. This is quite reliable, but it can be difficult to debug very obscure problems if the file that reveals the problems is automatically deleted. This doesn't require any additional acceptance test Then step to be sure the file is removed, since we don't need to test the `tempfile` module.

The cleaning application can also create a temporary file in the current working directory. This can be unlinked for normal operation, but left in place for debugging purposes. This will require at least two scenarios to be sure the file is removed normally, and be sure the file is retained to support debugging.

The final choice of implementation — and the related test scenarios — is left to you.

Project 3.6: Integration to create an acquisition pipeline

In *User experience*, we looked at the two-step user experience. One command is used to acquire data. After this, a second command is used to clean the data. An alternative user experience is a single shell pipeline.

Description

The previous projects in this chapter have decomposed the cleaning operation into two distinct steps. There's another, very desirable user experience alternative.

Specifically, we'd like the following to work, also:

```
% python src/acquire.py -s Series_1Pair --csv source.csv | python
src/clean.py -o analysis/Series_1.ndjson
```

The idea is to have the **acquire** application write a sequence of NDJSON objects to standard output. The **clean** application will read the sequence of NDJSON objects from standard input. The two applications will run concurrently, passing data from process to process.

For very large data sets, this can reduce the processing time. Because of the overhead in serializing Python objects to JSON text and deserializing Python objects from the text, the pipeline will not run in half the time of the two steps executed serially.

Multiple extractions

In the case of CSV extraction of Anscombe Quartet data, we have an **acquire** application that's capable of creating four files concurrently. This doesn't fit well with the shell pipeline. We have two architectural choices for handling this.

One choice is to implement a "fan-out" operation: the **acquire** program fans out data to four separate clean applications. This is difficult to express as a collection of shell pipelines. To implement this, a parent application uses concurrent.futures, queues, and processing pools. Additionally, the **acquire** program would need to write to shared queue objects, and the **clean** program would read from a shared queue.

The alternative is to process only one of the Anscombe series at a time. Introducing a -s Series_1Pair argument lets the user name a class that can extract a single series from the source data. Processing a single series at a time permits a pipeline that can be readily described as a shell command.

This concept is often necessary to disentangle enterprise data. It's common for enterprise applications — which often evolve organically — to have values from distinct problem domains as parts of a common record.

We'll turn to the technical approach in the next section.

Approach

For reference to the general approach to this application, see *Chapter 9, Project 3.1: Data Cleaning Base Application*, specifically *Approach*.

Writing the standard output (and reading from standard input) suggests that these applications will have two distinct operating modes:

- Opening a named file for output or input.

- Using an existing, open, unnamed file — often a pipe created by the shell — for output or input.

This suggests that the bulk of the design for an application needs to focus on open file-like objects. These are often described by the type hint of TextIO: they are files that can read (or write) text.

The top-level main() function must be designed either to open a named file, or to provide sys.stdout or sys.stdin as argument values. The various combinations of files are provided to a function that will do the more useful work.

This pattern looks like the following snippet:

```
if options.output:
    with options.output.open('w') as output:
```

```
        process(options.source, output)
else:
    process(options.source, sys.stdout)
```

The process() function is either given a file opened by a context manager, or the function is given the already open sys.stdout.

> The ability for a Python application to be part of a shell pipeline is a significant help in creating larger, more sophisticated composite processes. This higher-level design effort is sometimes called "Programming In The Large."
>
> Being able to read and write from pipelines was a core design feature of the Unix operating system and continues to be central to all of the various GNU/Linux variants.

This pipeline-aware design has the advantage of being slightly easier to unit test. The process() function's output argument value can be an io.StringIO object. When using a StringIO object, the file processing is simulated entirely in memory, leading to faster, and possibly simpler, tests.

This project sets the stage for a future project. See *Chapter 12, Project 3.8: Integrated Data Acquisition Web Service* for a web service that can leverage this pipeline.

Consider packages to help create a pipeline

A Python application to create a shell pipeline can involve a fair amount of programming to create two subprocesses that share a common buffer. This is handled elegantly by the shell.

An alternative is https://cgarciae.github.io/pypeln/. The **PypeLn** package is an example of a package that wraps the subprocess module to make it easier for a parent application to create a pipeline that executes the two child applications: **acquire** and **clean**.

Using a higher-level Python application to start the acquire-to-clean pipeline avoids the potential pitfalls of shell programming. It permits Python programs with excellent logging and debugging capabilities.

Now that we've seen the technical approach, it's appropriate to review the deliverables.

Deliverables

This project has the following deliverables:

- Documentation in the `docs` folder.

- Acceptance tests in the `tests/features` and `tests/steps` folders.

- Unit tests for the application modules in the `tests` folder.

- Revised applications that can be processed as a pipeline of two concurrent processes.

Many of these deliverables are described in previous chapters. Specifically, *Chapter 9, Project 3.1: Data Cleaning Base Application* covers the basics of the deliverables for this project.

Acceptance test

The acceptance test suite needs to confirm the two applications can be used as stand-alone commands, as well as used in a pipeline. One technique for confirming the pipeline behavior is to use shell programs like `cat` to provide input that mocks the input from another application.

For example, the `When` step may execute the following kind of command:

```
cat some_mock_file.ndj | python src/clean.py -o analysis/some_file.ndj
```

The **clean** application is executed in a context where it is part of an overall pipeline. The head of the pipeline is not the **acquire** application; we've used the `cat some_mock_file.ndj` command as a useful mock for the other application's output. This technique permits a lot of flexibility to test applications in a variety of shell contexts.

Using a pipeline can permit some helpful debugging because it disentangles two complicated programs into two smaller programs. The programs can be built, tested, and debugged in isolation.

Summary

This chapter expanded in several ways on the project in *Chapter 9, Project 3.1: Data Cleaning Base Application*. The following additional processing features were added:

- Pythonic approaches to validation and conversion of cardinal values.

- Approaches to validation and conversion of nominal and ordinal values.

- Techniques for uncovering key relationships and validating data that must properly reference a foreign key.

- Pipeline architectures using the shell pipeline.

Extras

Here are some ideas for you to add to these projects.

Hypothesis testing

The computations for mean, variance, standard deviation, and standardized Z-scores involve floating-point values. In some cases, the ordinary truncation errors of float values can introduce significant numeric instability. For the most part, the choice of a proper algorithm can ensure results are useful.

In addition to basic algorithm design, additional testing is sometimes helpful. For numeric algorithms, the **Hypothesis** package is particularly helpful. See `https://hypothesis.r eadthedocs.io/en/latest/`.

Looking specifically at *Project 3.5: Standardize data to common codes and ranges*, the *Approach* section suggests a way to compute the variance. This class definition is an excellent example of a design that can be tested effectively by the Hypothesis module to confirm that the results of providing a sequence of three known values produces the expected results for the count, sum, mean, variance, and standard deviation.

Rejecting bad data via filtering (instead of logging)

In the examples throughout this chapter, there's been no in-depth mention of what to do with data that raises an exception because it cannot be processed. There are three common choices:

1. Allow the exception to stop processing.

2. Log each problem row as it is encountered, discarding it from the output.

3. Write the faulty data to a separate output file so it can be examined with a data inspection notebook.

The first option is rather drastic. This is useful in some data cleaning applications where there's a reasonable expectation of very clean, properly curated data. In some enterprise applications, this is a sensible assumption, and invalid data is the cause for crashing the application and sorting out the problems.

The second option has the advantage of simplicity. A `try:`/`except:` block can be used to write log entries for faulty data. If the volume of problems is small, then locating the problems in the log and resolving them may be appropriate.

The third option is often used when there is a large volume of questionable or bad data. The rows are written to a file for further study.

You are encouraged to implement this third strategy: create a separate output file for rejected samples. This means creating acceptance tests for files that will lead to the rejection of at least one faulty row.

Disjoint subentities

An even more complicated data validation problem occurs when the source documents don't reflect a single resulting dataclass. This often happens when disjoint subtypes are merged into a single data set. This kind of data is a union of the disjoint types. The data must involve a "discriminator" field that shows which type of object is being described.

For example, we may have a few fields with date, time, and document ID that are common to all samples. In addition to those fields, a `document_type` field provides a set of codes to discriminate between the different kinds of invoices and different kinds of payments.

In this case, a conversion function involves two stages of conversions:

- Identify the subtype. This may involve converting the common fields and the discriminator field. The work will be delegated to a subtype-specific conversion for the rest of the work.

- Convert each subtype. This may involve a family of functions associated with each of the discriminator values.

This leads to a function design as shown in the activity diagram in *Figure 10.1*.

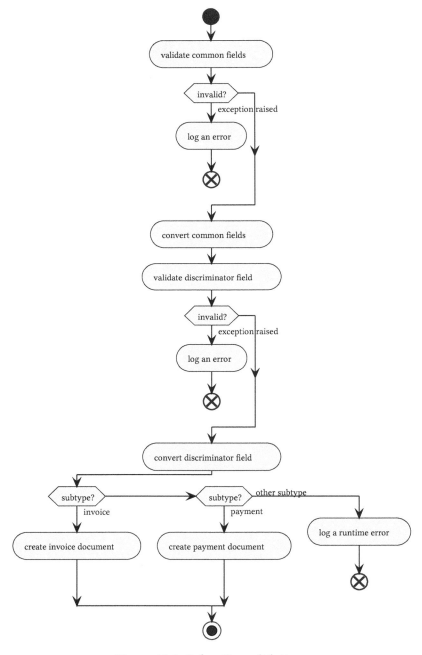

Figure 10.1: Subentity validation

Create a fan-out cleaning pipeline

There are two common alternatives for concurrent processing of a **acquire** and **clean** application:

- A shell pipeline that connects the **acquire** application to the **clean** application. These two subprocesses run concurrently. Each ND JSON line written by the **acquire** application is immediately available for processing by the **clean** application.

- A pool of workers, managed by concurrent.futures. Each ND JSON line created by the **acquire** application is placed in a queue for one of the workers to consume.

The shell pipeline is shown in *Figure 10.2*.

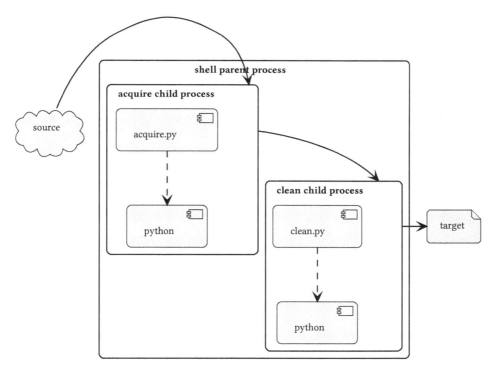

Figure 10.2: Components of a shell pipeline

The shell creates two child process with a shared buffer between them. For the acquire child process, the shared buffer is sys.stdout. For the clean child process, the shared buffer is sys.stdin. As the two applications run, each byte written is available to be read.

We've included explicit references to the Python runtime in these diagrams. This can help clarify how our application is part of the overall Python environment.

The pipeline creation is an elegant feature of the shell, and can be used to create complex sequences of concurrent processing. This is a handy way to think of decomposing a large collection of transformations into a number of concurrent transformations.

In some cases, the pipeline model isn't ideal. This is often the case when we need asymmetric collections of workers. For example, when one process is dramatically faster than another, it helps to have multiple copies of the slow processes to keep up with the faster process. This is handled politely by the concurrent.futures package, which lets an application create a "pool" of workers.

The pool can be threads or processes, depending on the nature of the work. For the most part, CPU cores tend to be used better by process pools, because OS scheduling is often process-focused. The Python **Global Interpreter Lock (GIL)** often prohibits compute-intensive thread pools from making effective use of CPU resources.

For huge data sets, worker-pool architecture can provide some performance improvements. There is overhead in serializing and deserializing the Python objects to pass the values from process to process. This overhead imposes some limitations on the benefits of multiprocessing.

The components that implement a worker process pool are shown in *Figure 10.3*.

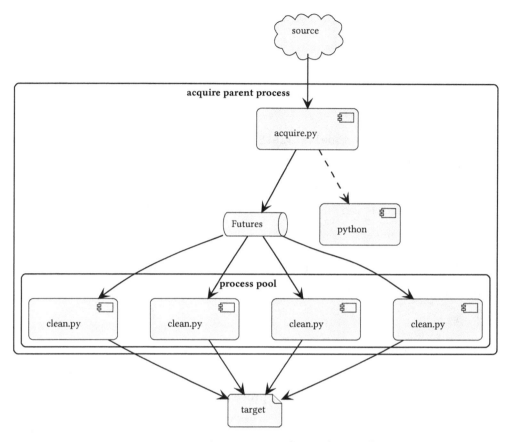

Figure 10.3: Components of a worker pool

This design is a significant alteration to the relationship between the acquire.py and clean.py applications. When the acquire.py application creates the process pool, it uses class and function definitions available within the same parent process.

This suggests the clean.py module needs to have a function that processes exactly one source document. This function may be as simple as the following:

```
from multiprocessing import get_logger


import acquire_model
import analysis_model
```

```
def clean_sample(
        acquired: acquire_model.SeriesSample
) -> analysis_model.SeriesSample:
    try:
        return analysis_model.SeriesSample.from_acquire_dataclass(acquired)
    except ValueError as ex:
        logger = get_logger()
        logger.error("Bad sample: %r\n%r\n", acquired, ex)
        return None
```

This function uses the analysis model definition, SeriesSample, to perform the validation, cleaning, and conversion of the acquired data. This can raise exceptions, which need to be logged.

The child processes are created with copies of the parent application's logging configuration. The multiprocessing.get_logger() function will retrieve the logger that was initialized into the process when the pool of worker processes was created.

The acquire.py application can use a higher-order map() function to allocate requests to the workers in an executor pool. The general approach is shown in the following incomplete code snippet:

```
with target_path.open('w') as target_file:
    with concurrent.futures.ProcessPoolExecutor() as executor:
        with source_path.open() as source:
            acquire_document_iter = get_series(
                source, builder
            )
            clean_samples = executor.map(
                clean.clean_sample,
                acquire_document_iter
```

```
        )
        count = clean.persist_samples(target_file, clean_samples)
```

This works by allocating a number of resources, starting with the target file to be written, then the pool of processes to write clean data records to the file, and finally, the source for the original, raw data samples. Each of these has a context manager to be sure the resources are properly released when all of the processing has finished.

We use the `ProcessPoolExecutor` object as a context manager to make sure the subprocesses are properly cleaned up when the source data has been fully consumed by the `map()` function, and all of the results retrieved from the `Future` objects that were created.

The `get_series()` function is an iterator that provides the builds the acquire version of each `SeriesSample` object. This will use an appropriately configured `Extractor` object to read a source and extract a series from it.

Since generators are lazy, nothing really happens until the values of the `acquire_document_iter` variable are consumed. The `executor.map()` will consume the source, providing each document to the pool of workers to create a `Future` object that reflects the work being done by a separate subprocess. When the work by the subprocess finishes, the `Future` object will have the result and be ready for another request.

When the `persist_samples()` functions consume the values from the `clean_samples` iterator, each of the `Future` objects will yield their result. These result objects are the values computed by the `clean.clean_sample()` function. The sequence of results is written to the target file.

The `concurrent.futures` process pool `map()` algorithm will preserve the original order. The process pool offers alternative methods that can make results ready as soon as they're computed. This can reorder the results; something that may or may not be relevant for subsequent processing.

11

Project 3.7: Interim Data Persistence

Our goal is to create files of clean, converted data we can then use for further analysis. To an extent, the goal of creating a file of clean data has been a part of all of the previous chapters. We've avoided looking deeply at the interim results of acquisition and cleaning. This chapter formalizes some of the processing that was quietly assumed in those earlier chapters. In this chapter, we'll look more closely at two topics:

- File formats and data persistence

- The architecture of applications

Description

In the previous chapters, particularly those starting with *Chapter 9, Project 3.1: Data Cleaning Base Application*, the question of "persistence" was dealt with casually. The previous chapters all wrote the cleaned samples into a file in ND JSON format. This saved delving into the alternatives and the various choices available. It's time to review the previous projects and consider the choice of file format for persistence.

What's important is the overall flow of data from acquisition to analysis. The conceptual flow of data is shown in *Figure 11.1*.

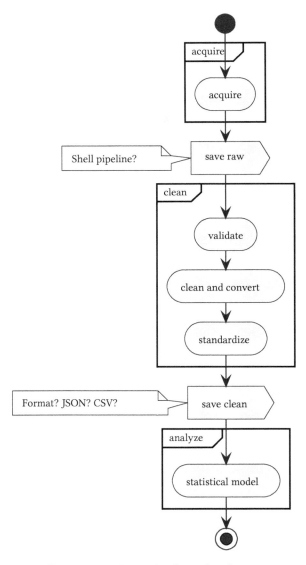

Figure 11.1: Data Analysis Pipeline

This differs from the diagram shown in *Chapter 2, Overview of the Projects*, where the stages were not quite as well defined. Some experience with acquiring and cleaning data helps to clarify the considerations around saving and working with data.

The diagram shows a few of the many choices for persisting interim data. A more complete list of format choices includes the following:

- CSV

- TOML

- JSON

- Pickle

- A SQL database

- YAML

There are others, but this list contains formats that enjoy a direct implementation in Python. Note that YAML is popular but isn't a built-in feature of the Python standard library. Additional formats include protocol buffers (`https://protobuf.dev`) and Parquet (`https://parquet.apache.org`). These two formats require a bit more work to define the structure before serializing and deserializing Python data; we'll leave them out of this discussion.

The CSV format has two disadvantages. The most notable of these problems is the representation of all data types as simple strings. This means any type of conversion information must be offered in metadata outside the CSV file. The **Pydantic** package provides the needed metadata in the form of a class definition, making this format tolerable. The secondary problem is the lack of a deeper structure to the data. This forces the files to have a flat sequence of primitive attributes.

The JSON format doesn't—directly—serialize datetime or timedelta objects. To make this work reliably, additional metadata is required to deserialize these types from supported JSON values like text or numbers. This missing feature is provided by the **Pydantic** package and works elegantly. A `datetime.datetime` object will serialize as a string, and the type information in the class definition is used to properly parse the string. Similarly, a `datetime.timedelta` is serialized as a float number but converted — correctly — into a `datetime.timedelta` based on the type information in the class definition.

The TOML format has one advantage over the JSON format. Specifically, the TOML format has a tidy way to serialize datetime objects, a capability the JSON library lacks. The TOML format has the disadvantage of not offering a direct way to put multiple TOML documents into a single file. This limits TOML's ability to handle vast datasets. Using a TOML file with a simple array of values limits the application to the amount of data that can fit into memory.

The pickle format can be used with the **Pydantic** package. This format has the advantage of preserving all of the Python-type information and is also very compact. Unlike JSON, CSV, or TOML, it's not human-friendly and can be difficult to read. The `shelve` module permits the building of a handy database file with multiple pickled objects that can be saved and reused. While it's technically possible to execute arbitrary code when reading a pickle file, the pipeline of acquisition and cleansing applications does not involve any unknown agencies providing data of unknown provenance.

A SQL database is also supported by the **Pydantic** package by using an ORM model. This means defining two models in parallel. One model is for the ORM layer (for example, SQLAlchemy) to create table definitions. The other model, a subclass of `pydantic.BaseModel`, uses native **Pydantic** features. The **Pydantic** class will have a `from_orm()` method to create native objects from the ORM layer, performing validation and cleaning.

The YAML format offers the ability to serialize arbitrary Python objects, a capability that makes it easy to persist native Python objects. It also raises security questions. If care is taken to avoid working with uploaded YAML files from insecure sources, the ability to serialize arbitrary Python code is less of a potential security problem.

Of these file formats, the richest set of capabilities seems to be available via JSON. Since we'll often want to record many individual samples in a single file, **newline-delimited (ND)** JSON seems to be ideal.

In some situations — particularly where spreadsheets will be used for analysis purposes — the CSV format offers some value. The idea of moving from a sophisticated Jupyter Notebook to a spreadsheet is not something we endorse. The lack of automated test capabilities for spreadsheets suggests they are not suitable for automated data processing.

Overall approach

For reference see *Chapter 9, Project 3.1: Data Cleaning Base Application*, specifically *Approach*. This suggests that the `clean` module should have minimal changes from the earlier version.

A cleaning application will have several separate views of the data. There are at least four viewpoints:

- The source data. This is the original data as managed by the upstream applications. In an enterprise context, this may be a transactional database with business records that are precious and part of day-to-day operations. The data model reflects considerations of those day-to-day operations.

- Data acquisition interim data, usually in a text-centric format. We've suggested using ND JSON for this because it allows a tidy dictionary-like collection of name-value pairs, and supports quite complex Python data structures. In some cases, we may perform some summarization of this raw data to standardize scores. This data may be used to diagnose and debug problems with upstream sources. It's also possible that this data only exists in a shared buffer as part of a pipeline between an acquire and a cleaning application.

- Cleaned analysis data, using native Python data types including `datetime`, `timedelta`, `int`, `float`, and `boolean`. These are supplemented with **Pydantic** class definitions that act as metadata for proper interpretation of the values. These will be used by people to support decision-making. They may be used to train AI models used to automate some decision-making.

- The decision-maker's understanding of the available information. This viewpoint tends to dominate discussions with users when trying to gather, organize, and present data. In many cases, the user's understanding grows and adapts quickly as data is presented, leading to a shifting landscape of needs. This requires a great deal of flexibility to provide the right data to the right person at the right time.

The **acquire** application overlaps with two of these models: it consumes the source data and produces an interim representation. The **clean** application also overlaps two of these models: it consumes the interim representation and produces the analysis model objects. It's essential to distinguish these models and to use explicit, formal mappings between them.

This need for a clear separation and obvious mappings is the primary reason why we suggest including a "builder" method in a model class. Often we've called it something like `from_row()` or `from_dict()` or something that suggests the model instance is built from some other source of data via explicit assignment of individual attributes.

Conceptually, each model has a pattern similar to the one shown in the following snippet:

```python
class Example:
        field_1: SomeElementType
        field_2: AnotherType


        @classmethod
        def from_source(cls, source: SomeRowType) -> "Example":
                return Example(
                        field_1=transform_1(source),
                        field_2=transform_2(source),
                )
```

The transformation functions, `transform1()` and `transform2()`, are often implicit when using `pydantic.BaseModel`. This is a helpful simplification of this design pattern. The essential idea, however, doesn't change, since we're often rearranging, combining, and splitting source fields to create useful data.

When the final output format is either CSV or JSON, there are two helpful methods of
`pydantic.BaseModel`. These methods are `dict()` and `json()`. The `dict()` method creates
a native Python dictionary that can be used by a `csv.DictWriter` instance to write CSV
output. The `json()` method can be used directly to write data in ND JSON format. It's
imperative for ND JSON to make sure the `indent` value used by the `json.dump()` function
is `None`. Any other value for the `indent` parameter will create multi-line JSON objects,
breaking the ND JSON file format.

The **acquire** application often has to wrestle with the complication of data sources that
are unreliable. The application should save some history from each attempt to acquire data
and acquire only the "missing" data, avoiding the overhead of rereading perfectly good
data. This can become complicated if there's no easy way to make a request for a subset
of data.

When working with APIs, for example, there's a `Last-Modified` header that can help
identify new data. The `If-Modified-Since` header on a request can avoid reading data
that's unchanged. Similarly, the `Range` header might be supported by an API to permit
retrieving parts of a document after a connection is dropped.

When working with SQL databases, some variants of the `SELECT` statement permit `LIMIT`
and `OFFSET` clauses to retrieve data on separate pages. Tracking the pages of data can
simplify restarting a long-running query.

Similarly, the **clean** application needs to avoid re-processing data in the unlikely event
that it doesn't finish and needs to be restarted. For very large datasets, this might mean
scanning the previous, incomplete output to determine where to begin cleaning raw data
to avoid re-processing rows.

We can think of these operations as being "idempotent" in the cases when they have run
completely and correctly. We want to be able to run (and re-run) the **acquire** application
without damaging intermediate result files. We also want an additional feature of adding to
the file until it's correct and complete. (This isn't precisely the definition of "idempotent";

we should limit the term to illustrate that correct and complete files are not damaged by re-running an application.) Similarly, the **clean** application should be designed so it can be run — and re-run — until all problems are resolved without overwriting or reprocessing useful results.

Designing idempotent operations

Ideally, our applications present a UX that can be summarized as "pick up where they left off." The application will check for output files, and avoid destroying previously acquired or cleaned data.

For many of the carefully curated Kaggle data sets, there will be no change to the source data. A time-consuming download can be avoided by examining metadata via the Kaggle API to decide if a file previously downloaded is complete and still valid.

For enterprise data, in a constant state of flux, the processing must have an explicit "as-of date" or "operational date" provided as a run-time parameter. A common way to make this date (or date-and-time) evident is to make it part of a file's metadata. The most visible location is the file's name. We might have a file named `2023-12-31-manufacturing-orders.ndj`, where the as-of date is clearly part of the file name.

Idempotency requires programs in the data acquisition and cleaning pipeline to check for existing output files and avoid overwriting them unless an explicit command-line option permits overwriting. It also requires an application to read through the output file to find out how many rows it contains. This number of existing rows can be used to tailor the processing to avoid re-processing existing rows.

Consider an application that reads from a database to acquire raw data. The "as-of-date" is 2022-01-18, for example. When the application runs and something goes wrong in the network, the database connection could be lost after processing a subset of rows. We'll imagine the output file has 42 rows written before the network failure caused the application to crash.

When the log is checked and it's clear the application failed, it can be re-run. The program can check the output directory and find the file with 42 rows, meaning the application is being run in recovery mode. There should be two important changes to behavior:

- Add a `LIMIT -1 OFFSET 42` clause to the `SELECT` statement to skip the 42 rows already retrieved. (For many databases, `LIMIT -1 OFFSET 0` will retrieve all rows; this can be used as a default value.)

- Open the output file in "append" mode to add new records to the end of the existing file.

These two changes permit the application to be restarted as many times as required to query all of the required data.

For other data sources, there may not be a simple "limit-offset" parameter in the query. This may lead to an application that reads and ignores some number of records before processing the remaining records. When the output file doesn't exist, the offset before processing has a value of zero.

It's important to handle date-time ranges correctly.

It's imperative to make sure date and date-time ranges are properly **half-open intervals**. The starting date and time are included. The ending date and time are excluded.

Consider a weekly extract of data.

One range is 2023-01-14 to 2023-01-21. The 14th is included. The 21st is not included. The next week, the range is 2023-01-21 to 2023-01-28. The 21st is included in this extract.

Using half-open intervals makes it easier to be sure no date is accidentally omitted or duplicated.

Now that we've considered the approach to writing the interim data, we can look at the deliverables for this project.

Deliverables

The refactoring of existing applications to formalize the interim file formats leads to changes in existing projects. These changes will ripple through to unit test changes. There should not be any acceptance test changes when refactoring the data model modules.

Adding a "pick up where you left off" feature, on the other hand, will lead to changes in the application behavior. This will be reflected in the acceptance test suite, as well as unit tests.

The deliverables depend on which projects you've completed, and which modules need revision. We'll look at some of the considerations for these deliverables.

Unit test

A function that creates an output file will need to have test cases with two distinct fixtures. One fixture will have a version of the output file, and the other fixture will have no output file. These fixtures can be built on top of the pytest.tmp_path fixture. This fixture provides a unique temporary directory that can be populated with files needed to confirm that existing files are appended to instead of overwritten.

Some test cases will need to confirm that existing files were properly extended. Other test cases will confirm that the file is properly created when it didn't exist. An edge case is the presence of a file of length zero — it was created, but no data was written. This can be challenging when there is no previous data to read to discover the previous state.

Another edge case is the presence of a damaged, incomplete row of data at the end of the file. This requires some clever use of the seek() and tell() methods of an open file to selectively overwrite the incomplete final record of the file. One approach is to use the tell() method before reading each sample. If an exception is raised by the file's parser, seek to the last reported tell() position, and start writing there.

Acceptance test

The acceptance test scenarios will require an unreliable source of data. Looking back at *Chapter 4, Data Acquisition Features: Web APIs and Scraping*, specifically *Acceptance tests*, we can see the acceptance test suite involves using the bottle project to create a very small web service.

There are two aspects to the scenarios, each with different outcomes. The two aspects are:

1. The service or database provides all results or it fails to provide a complete set of results.

2. The working files are not present — we could call this the "clean start" mode — or partial files exist and the application is working in recovery mode.

Since each aspect has two alternatives, there are four combinations of scenarios for this feature:

1. The existing scenario is where the working directory is empty and the API or database works correctly. All rows are properly saved.

2. A new scenario where the working directory is empty and the service or database returns a partial result. The returned rows are saved, but the results are marked as incomplete, perhaps with an error entry in the log.

3. A new scenario where the given working directory has partial results and the API or database works correctly. The new rows are appended to existing rows, leading to a complete result.

4. A new scenario where the given working directory has partial results and the service or database returns a partial result. The cumulative collection of rows are usable, but the results are still marked as incomplete.

A version of the mock RESTful process can return some rows and even after that return 502 status codes. The database version of the incomplete results scenarios is challenging because SQLite is quite difficult to crash at run-time. Rather than try to create a version of

SQLite that times out or crashes, it's better to rely on unit testing with a mock database to be sure crashes are handled properly. The four acceptance test scenarios will demonstrate that working files are extended without being overwritten.

Cleaned up re-runnable application design

The final application with the "pick-up-where-you-left-off" feature can be very handy for creating robust, reliable analytic tools. The question of "what do we do to recover?" should involve little (or no) thought.

Creating "idempotent" applications, in general, permits rugged and reliable processing. When an application doesn't work, the root cause must be found and fixed, and the application can be run again to finish the otherwise unfinished work from the failed attempt. This lets analysts focus on what went wrong — and fixing that — instead of having to figure out how to finish the processing.

Summary

In this chapter, we looked at two important parts of the data acquisition pipeline:

- File formats and data persistence

- The architecture of applications

There are many file formats available for Python data. It seems like newline delimited (ND) JSON is, perhaps, the best way to handle large files of complex records. It fits well with Pydantic's capabilities, and the data can be processed readily by Jupyter Notebook applications.

The capability to retry a failed operation without losing existing data can be helpful when working with large data extractions and slow processing. It can be very helpful to be able to re-run the data acquisition without having to wait while previously processed data is processed again.

Extras

Here are some ideas for you to add to these projects.

Using a SQL database

Using a SQL database for cleaned analytical data can be part of a comprehensive database-centric data warehouse. The implementation, when based on **Pydantic**, requires the native Python classes as well as the ORM classes that map to the database.

It also requires some care in handling repeated queries for enterprise data. In the ordinary file system, file names can have processing dates. In the database, this is more commonly assigned to an attribute of the data. This means multiple time periods of data occupy a single table, distinguished by the "as-of" date for the rows.

A common database optimization is to provide a "time dimension" table. For each date, the associated date of the week, fiscal weeks, month, quarter, and year is provided as an attribute. Using this table saves computing any attributes of a date. It also allows the enterprise fiscal calendar to be used to make sure that 13-week quarters are used properly, instead of the fairly arbitrary calendar month boundaries.

This kind of additional processing isn't required but must be considered when thinking about using a relational database for analysis data.

This extra project can use SQLAlchemy to define an ORM layer for a SQLite database. The ORM layer can be used to create tables and write rows of analysis data to those tables. This permits using SQL queries to examine the analysis data, and possibly use complex SELECT-GROUP BY queries to perform some analytic processing.

Persistence with NoSQL databases

There are many NoSQL databases available. A number of products like MongoDB use a JSON-based document store. Database engines like PostgreSQL and SQLite3 have the capability of storing JSON text in a column of a database table. We'll narrow our focus onto JSON-based databases as a way to avoid looking at the vast number of databases available.

We can use SQLite3 BLOB columns to store JSON text, creating a NoSQL database using the SQLite3 storage engine.

A small table with two columns: doc_id, and doc_text, can create a NoSQL-like database. The SQL definition would look like this:

```
CREATE TABLE IF NOT EXISTS document(
        doc_id INTEGER PRIMARY KEY,
        doc_text BLOB
)
```

This table will have a primary key column that's populated automatically with integer values. It has a text field that can hold the serialized text of a JSON document.

The SQLite3 function json() should be used when inserting JSON documents:

```
INSERT INTO document(doc_text) VALUES(json(:json_text))
```

This will confirm the supplied value of json_text is valid JSON, and will also minimize the storage, removing needless whitespace. This statement is generally executed with the parameter {"json_text": json.dumps(document) to convert a native Python document into JSON text so it can then be persisted into the database.

The attributes of a JSON object can be interrogated using the SQLite ->> operator to extract a field from a JSON document. A query for a document with a named field that has a specific value will look like this:

```
SELECT doc_text FROM document WHERE doc_text ->> 'field' = :value
```

In the above SQL, the field's name, field, is fixed as part of the SQL. This can be done when the schema is designed to support only a few queries. In the more general case, the field name might be provided as a parameter value, leading to a query like the following:

```
SELECT doc_text FROM document WHERE doc_text ->> :name = :value
```

This query requires a small dictionary with the keys "name" and "value", which will provide the field name and field value used to locate matching documents.

This kind of database design lets us write processing that's similar to some of the capabilities of a document store without the overhead of installing a document store database. The JSON documents can be inserted into this document store. The query syntax uses a few SQL keywords as overhead, but the bulk of the processing can be JSON-based interrogation of documents to locate the desired subset of available documents.

The idea here is to use a JSON-based document store instead of a file in ND JSON format. The Document Store interface to SQLite3 should be a module that can be reused in a JupyterLab Notebook to acquire and analyze data. While unit tests are required for the database interface, there are a few changes to the acceptance test suite required to confirm this changed design.

12

Project 3.8: Integrated Data Acquisition Web Service

In many enterprise applications, data is provided to several consumers. One way to do this is to define an API that provides data (and the metadata) for subsequent use. In this chapter, we guide you through the transformation of Project 2.5 schema information into a larger OpenAPI specification. We will also build a small Flask application that provides the core acquire-cleanse-convert process as a web service.

We'll cover a number of skills in the chapter:

- Creating an OpenAPI specification for a service to acquire and download data

- Writing a web service application to implement the OpenAPI specification

- Using a processing pool to delegate long-running background tasks

This is a bit of a deviation from a straight path of acquiring and cleaning data. In some enterprises, this deviation is needed to publish useful data to a wider audience.

We'll begin with a description of the behavior of this RESTful API server.

Description

In *Chapter 8, Project 2.5: Schema and Metadata*, we used **Pydantic** to generate a schema for the analysis data model. This schema provides a formal, language-independent definition of the available data. This can then be shared widely to describe the data and resolve questions or ambiguities about the data, the processing provenance, the meaning of coded values, internal relationships, and other topics.

This specification for the schema can be extended to create a complete specification for a RESTful API that provides the data that meets the schema. The purpose of this API is to allow multiple users — via the `requests` module — to query the API for the analytical data as well as the results of the analysis. This can help users to avoid working with out-of-date data. An organization creates large JupyterLab servers to facilitate doing analysis processing on machines far larger than an ordinary laptop.

Further, an API provides a handy wrapper around the entire acquire-and-clean process. When a user requests data for the first time, the processing steps can be started and the results cached. Each subsequent request can download available data from a filesystem cache, providing rapid access. In the case of a failure, the logs can be provided as an alternative to the final data.

We won't dive deeply into REST design concepts. For more information on RESTful design, see `https://hub.packtpub.com/creating-restful-api/`.

Generally, a RESTful API defines a number of paths to resources. A given path can be accessed by a number of methods, some of which will get the resource. Other methods may post, patch, put, or delete the resource. The defined HTTP methods offer handy mapping to the common **Create-Retrieve-Update-Delete** (**CRUD**) conceptual operations.

Here are the common cases:

- A path without a final identifier, for example, `/series/`. There are two common cases here:

- The GET method will retrieve the list of available resources of the given type.

- The POST method can be used to create a new instance of this type. This is the conceptual "Create" operation.

- A path with an identifier. For example, /series/Series_4. This is a specific resource. There are several methods that might be implemented:

  - The GET method will retrieve the resource. This is the "Retrieve" conceptual operation.

  - The PUT and PATCH methods can be used to replace or update the resource. These are two forms of the conceptual "Update" operation.

  - The DELETE method can be used to remove the resource. This is the "Delete" conceptual operation.

It becomes imperative to consider a RESTful web service as a collection of resources. Talking about resources can make it difficult to talk about a RESTful request that initiates processing. It raises the question of what resource describes an activity such as processing samples. We'll start by considering the data series as the most important resource provided by this service.

The data series resources

The primary resource for this API is the data series. As shown in the previous section, *OpenAPI 3 specification*, a path with /2023.02/series/<id> can be used to extract the data for a named series. The 2023.02 prefix allows the API to evolve to a newer version while leaving older paths in place for compatibility purposes.

The use of **semantic versioning (semver)** is common, and many APIs have something like "v1" in the path. Yet another alternative is to include the version information in the Accept header. This means the URIs never change, but the schema for the response can change based on the version information provided in the header.

The various "series" routes provide direct access to the data resources. This seems appropriate since this is the primary purpose of the service.

There is an additional class of resources that might be of interest: the background processing used to create the data. As noted above, projects like *Chapter 11, Project 3.7: Interim Data Persistence*, are the essential foundation for processing done by this RESTful API. The acquire and clean applications can be run in the background to create data for download.

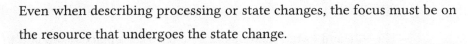

A focus on resources is essential for making useful RESTful APIs.

Even when describing processing or state changes, the focus must be on the resource that undergoes the state change.

The methods available in HTTP (GET, POST, PUT, PATCH, and DELETE, for example) are effectively the verbs of the API's language. The resources are nouns.

Creating data for download

The primary purpose of the RESTful API is to store and download clean data for analysis work. This can be a relatively straightforward application that offers data files from a well-known directory. The work involves matching RESTful requests against available files, and returning appropriate status codes when requests are made for files that don't exist.

A secondary purpose is to automate the creation of the data for download. The RESTful API can be a wrapper around the complete acquire, clean, and persist pipeline. To do this, the API will have two distinct kinds of requests:

- Requests to download existing, cached data. The resource type is clear here.

- Requests to start the creation of new data; this will lead to cached data available for download. The resource type for processing isn't as clear.

An operation or action does have some static resources that can be used with a RESTful API. Here are two common resource types for activities:

- A "current status" resource that reflects the work being done right now

- A "processing history" resource that reflects work completed: this is often the log file for the acquisition processing

The control of processing by a RESTful API can work by creating and examining processing status or history as a distinct resource type:

- A path with a POST request will start an asynchronous, background process. This will also create a new processing history resource. The response body provides a transaction identifier referring to this new processing history.

- A path with a transaction identifier and a GET request will return the background processing details; this should include the current or final status as well as the log.

For sophisticated frontend processing, a web socket can be created to receive ongoing status reports from the background process. For a less sophisticated frontend, polling every few seconds can be done to see whether the processing has finished and the data is available for download.

With both processing history resources and data resources, the following two sets of paths are necessary:

- `/series/<id>` paths that refer to specific series, already available in the cache. These resources are accessed exclusively with the GET method to download data.

- `/creation/<id>` paths that refer to background processing jobs to create a new series of data. These resources will use the POST method to start a background job, and the GET method to check the status of a job.

This set of paths (and the associated methods) allows a user to control processing and check the results of processing. The user can ask for available datasets and download a specific dataset for analysis.

Overall approach

We'll take some guidance from the C4 model (`https://c4model.com`) when looking at our approach.

- **Context** For this project, the context diagram has several use cases: listing available data, downloading available data, starting a process to acquire data, and checking the status of a process acquiring data.

- **Containers** Ideally, this runs on a single container that hosts the web service as well as the processing. In some cases, multiple containers will be required because the processing demands are so huge.

- **Components** There are two significantly different collections of software components: the web service, and the application programs that run in the background to acquire and clean the data.

- **Code** The acquiring and cleaning applications have already been described as separate projects. We'll focus on the web service.

We'll decompose the web service application into several components. The following diagram shows the relationship between the RESTful API service and the applications that are run to acquire and clean data.

The component diagram is shown in *Figure 12.1.*

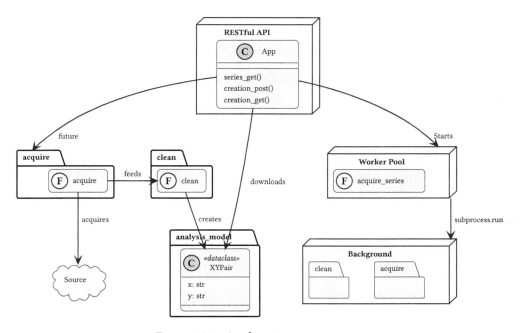

Figure 12.1: Application components

This diagram shows three separate processes:

- The **RESTful API** process that handles HTTP requests from clients.

- The **Worker Pool** collection of processes that are managed by the concurrent.futures module. Each worker will be running a single function, shown as acquire_series, that's defined in the same module as the **RESTful API** service.

- The **Background** process that is executed by a worker in the worker pool. This uses the subprocess module to run an existing CLI application.

When the API service starts, it uses concurrent.futures to create a pool of workers. A request to acquire and clean data will use the submit() method of the pool to create a **future**. This future is a reference to a subprocess that will — eventually — return the final status of the acquire and clean job. The subprocess that implements the future will evaluate the `acquire_series()` function defined in the same module as the RESTful API application to do the work.

When the `acquire_series()` function finishes the processing, it will have created a file that can be downloaded. Via the future object, it will also provide some status information to the RESTful API service to indicate the processing is done.

One suggested implementation for the `acquire_series()` function is to use `subprocess.run()` to execute the acquire and clean applications to gather and cleanse source data. There are some other choices available. The most important alternative is to import these two other modules, and execute them directly, rather than creating a subprocess. This direct execution has the advantage of being slightly faster than spawning a subprocess. It has the disadvantage of making it more complicated to create a separate log file each time the **acquire** and **clean** application is executed.

We'll take a look at the OpenAPI specification for the RESTful API first. This helps to characterize the overall UX.

OpenAPI 3 specification

A RESTful API requires a clear description of the requests and responses. The OpenAPI specification is a formal definition of RESTful web services. See `https://www.openapis.org`. This document has a version identifier and some information about the service as a whole. For this project, the most important part is the **paths** section, which lists the various resource types and the paths used to locate those resources. The **components** section provides the needed schema definitions.

An OpenAPI document often has an outline like this:

```
{
        "openapi": "3.0.3",
        "info": {
                "title": "The name of this service",
                "description": "Some details.",
```

```
            "version": "2023.02"
    }
    "paths": {
            "..."
    }
    "components": {
            "parameters": {"..."},
            "schemas": {"..."}
    }
}
```

The details of the paths and components have been elided from this overview. (We've used
"..." in place of the details.) The idea is to show the general structure of an OpenAPI
specification. While JSON is the underlying format commonly used for these specifications,
it can be hard to read. For this reason, it's common to use YAML notation for OpenAPI
specifications.

> Think of the OpenAPI specification as a binding contract.
>
> The acceptance test suite should be Gherkin scenarios with a very direct
> mapping to the OpenAPI specification.
>
> For more on the idea of OpenAPI to Gherkin, see `https://medium.com/c`
> `apital-one-tech/spec-to-gherkin-to-code-902e346bb9aa`.

The OpenAPI paths define the resources made available by a RESTful API. In this case, the
resources are cleaned files, ready for analysis.

We'll often see entries in the **paths** section that look like the following YAML snippet:

```
"/2023.02/series":
  get:
    responses:
```

```
    "200":
      description: All of the available data series.
      content:
        application/json:
          schema:
            $ref: "#/components/schemas/series_list"
```

This shows a path that starts with an API version number (in this example, calendar versioning, "calver", is used) and a resource-type, `series`. Any given path can be accessed by a variety of methods; in this example, only the **get** method is defined.

One kind of response is defined for requests to this path and method combination. The response will have a status code of 200, meaning normal, successful completion. The description is there to explain what this resource will be. A response can define a variety of content types; in this example, only `application/json` is defined. The schema for this is provided elsewhere in the OpenAPI specification, in the `components/schemas` section of the document.

The use of a `$ref` tag within the specification permits common definitions, such as schemas and parameters, to be collected under the `components` section, permitting reuse. This follows the **DRY (Don't Repeat Yourself)** principle of software design.

It can be difficult to get the syntax correct in an OpenAPI specification. It's helpful to have an editor that validates the specification. For example, `https://editor.swagger.io` provides an editor that helps confirm the specification is internally consistent. For readers using tools such as JetBrains' PyCharm, there's a plug-in editor: `https://plugins.jetb rains.com/plugin/14837-openapi-swagger-editor`.

When a path has an identifier in it, then this is shown with the path name of the form `"/2023.02/series/<series_id>"`. The `<series_id>` is defined in the `parameters` section of this request. Since parameters are sometimes reused, it's helpful to have a reference to a component with the common definition.

The whole request might start like this:

```
"/2023.02/series/<series_id>":
  get:
    description:
      Get series data as text ND JSON.
    parameters:
      - $ref:
          "#/components/parameters/series_id"
    responses:

      ...
```

The details of the **responses** section have been omitted from this example. The parameter definition — in the components section — might look like this:

```
series_id:
  name: series_id
  in: path
  required: true
  description: Series name.
  schema:
    type: string
```

This provides a wealth of details about the series_id parameter, including the description and a formal schema definition. For simple APIs, the name of the parameter and the reference label under components are often the same. In more complex cases, a parameter name might be reused, but have distinct semantics in distinct contexts. A generic word such as id might be used in several different paths, leading to the reference label being something more descriptive than id.

The content for ND JSON is considered an extension to standard MIME types. Therefore the content definition for a response that includes data might look like this:

```
content:
  application/x-ndjson:
```

```
schema:
    $ref: "#/components/schemas/samples"
```

The schema is a challenge because it pushes the boundaries of what JSON Schema can describe. It looks as follows:

```
samples:
    description: >
        Acquired data for a series in ND JSON format.
        See http://ndjson.org and https://jsonlines.org.
    type: string
    format: "(\\{.*?\\}\\n)+"
```

The format information describes the physical organization of ND JSON data, but doesn't provide any details on the structure of the schema for each individual row. The additional schema details can either be added to the description, or a separate label, distinct from other JSON schema labels, can be used, for example, "ndjson-schema:".

RESTful API to be queried from a notebook

The RESTful API service must be a wrapper around application programming that can perform the required processing. The idea is to put as little processing as possible into the RESTful API. It serves as a very thin — almost transparent — interface to the "real work" of the application. For this reason, projects such as *Chapter 11, Project 3.7: Interim Data Persistence* are the essential foundation of this RESTful API.

As noted in *Figure 12.1*, the **Background** processing is completely outside the RESTful API. This separation of concerns is absolutely essential. The general processing of samples can be performed with a CLI or through the RESTful API and create identical results.

If additional processing — for example, additional cleaning — is done by the RESTful service, then there are results that can't be reproduced from the CLI. This means the acceptance test suites have distinct results. This will lead to problems when a change is made to the underlying **acquire** or **clean** application and the "extra" processing that was jammed into the RESTful service now appears to be broken.

A common source of problems in enterprise software is the failure to honor the **Interface Segregation** design principle. A complex application may be supported by several collaborating organizations. When one organization is slow to respond to requests for changes, another organization may step in and make a bad design decision, implementing processing in the API interface that should have been part of a background module with a proper CLI interface. The urge to be responsive to customers can often overshadow the importance of the separation of concerns.

For this project, the server can be built as a single process, avoiding the need for the distributed cache. Further, because the data series and the processing logs are all simple files, a database is not required; the local filesystem is perfectly suited to this service.

To create a more scalable solution, a library such as **celery** can be used to create a more robust distributed worker pool. This isn't needed for a small server, however.

In the next section, we'll review how processing can be started by a RESTful API.

A POST request starts processing

The general approach to creating a new resource is to make a POST request to a path. This will either return a 400 error status or it will issue a redirect (301) to a new path to retrieve the status of the background processing. This pattern is called the **Post-Redirect-Get** design pattern. It permits a user interacting with a browser to use the **back** button to perform the GET method again; it prevents the **back** button from submitting a duplicate request.

For a client application making a request via requests the redirect is essentially invisible. The request history will reveal the redirection. Also, the full URL recorded in the response will reflect the redirection.

The general processing for this route, then, is as follows:

1. Validate all of the parameters to make sure they describe the series and the source of the data. If there is anything amiss, a JSON response with the details of the problem

must be returned, with a status code of 400 to indicate the request is invalid and must be changed.

2. Use the worker pool `submit()` method to create a `Future` object. This object can be saved in a local cache by the RESTful API. This cache of `Future` objects can be queried to see the background processing currently being performed. The future's result is usually something indicative of success or failure; for example, the return code from the subprocess – usually a zero indicates success.

3. Use the `redirect()` function in the Bottle framework to return the status code to direct a client to another URL for the status of the just-created `Future` object. Separately, a GET request will prepare a JSON document with the status of the job creating the data.

When using a framework like Bottle, this function is marked with a `@post("/2023.02/creation")` decorator. This names the POST method and the path that will be handled by the function.

The log files from processing can be the longer-term repository of processing history. The GET request for status will return a log and possibly the state of an active `Future` object. We'll look at this request next.

The GET request for processing status

The initial POST request to start processing will redirect to a GET request that reveals the status of the processing. The initial response may have almost no other details beyond the fact that the processing job has started.

This status path should return one of two things:

- A 404 status if the process ID is unknown. This would mean no previous request had been made with this identifier and no current request has this identifier, either.

- A 200 status with JSON content that includes some combination of two things: the state of a future object and the log file.

Most users will only care about the state of the `Future` object. In the case of developers, however, who are adding features to data acquire or data cleaning applications, then the log might be important support for observability.

When using a framework like Bottle, this function is marked with a `@get("/2023.02/creation/<job_id>")` decorator. This provides the method and the path that will be handled by the function. The use of `<job_id>` parses this section of the path and provides the value as a separate parameter to the function that implements this route.

Once the processing is complete, a subsequent request can provide the data. We'll look at this next.

The GET request for the results

This path should return one of two things:

- A 404 status if the series identifier is unknown.

- A 200 status with the ND JSON content. This has a MIME type of `application/x-ndjson` to indicate it's an extension to the standard collection of MIME types.

When using a framework like Bottle, this function is marked with a `@get("/2023.02/series/<series_id>")` decorator. The use of `<series_id>` parses this section of the path and provides the value as a separate parameter to the function that implements this route.

A more sophisticated implementation can check for an `Accept` header in the request. This header will state the preferred MIME type, and might have `text/csv` instead of `application/x-ndjson`. The use of this header permits a client to make requests for data in a format the application finds most useful.

Security considerations

A RESTful API requires some care to be sure the requests fit with the overall enterprise information access policies. In some cases, this might mean individual access controls to be sure each person can access permitted data. There are numerous **Single Sign-On (SSO)** products that can handle the identity of individuals.

Another common approach is to have an API work with assigned API keys. The team supporting the API can provide unique API key values to known users or teams. Within most enterprises, there's little need for automating the assignment of API keys for internal-facing APIs. The set of valid API keys may be reduced or expanded to reflect organizational merges and splits.

 API key values are sent from the client to the server to authenticate the user making a request. They are never sent from the server to a client. The API keys can be kept in a simple text file; the file's permissions should restrict it to read-only access by the account handling the service as a whole. This requires administrators to take steps to manage the file of API keys to avoid damaging it or revealing it to unauthorized users.

When working with API keys, there are a number of ways the client can provide the key with each API request. One of the more popular techniques is to use these complementary security features:

- The HTTPS protocol, where all of the communication between client and server application is encrypted.

- The HTTP **Authorization** header with **Basic** authorization. This header will have a username and the API key as the password.

The use of the **Authorization** header is often very simple for a client tool. Many libraries — for example, the **requests** library — offer an object class that contains the username and API key. Using the auth= parameter on a request function will build the appropriate header.

The use of HTTPS includes **Transport Layer Security (TLS)** to keep the content of the **Authorization** header secret. The **requests** package handles this politely.

On the server side, each of these must be handled by our RESTful API application. Using HTTPS is best done by running the **Bottle** application inside another server. We could, for example, create an NGINX and uWSGI configuration that would run our RESTful app inside a containing server. Another choice is to use a Python-based server such as Paste or GUnicorn to contain the **Bottle** application. It's essential to have a container server to handle the details of HTTPS negotiation.

Processing the **Authorization** header is something best done within the RESTful API. Some routes (i.e., the openapi.yaml) should not include any security considerations. Other routes — specifically those that cause state changes — may be limited to a subset of all users.

This suggests the list of users includes some permissions as well as their API key. Each route needs to confirm the **Authorization** header has a known user and the correct key. The request.auth property of the request object is a two-tuple with the username and API key value. This can be used to decide whether the request is generally acceptable, and also to decide whether a state-changing **Post** operation is permitted for the given user. This kind of processing is often implemented as a decorator.

We won't dig deeply into the design of this decorator. For this project, with so few resources, a repeated if statement inside each function is acceptable.

Deliverables

This project has the following deliverables:

- Documentation in the docs folder

- Acceptance tests in the tests/features and tests/steps folders

- Unit tests for the application modules in the tests folder

- An application for the RESTful API processing

We'll start by looking at the acceptance test cases, first. They'll be rather complex because we need to start the RESTful API service before we can access it with a client request.

Acceptance test cases

Back in *Chapter 4, Data Acquisition Features: Web APIs and Scraping*, specifically *Acceptance tests using a SQLite database*, we looked at ways to describe a scenario that involved a database service.

For this project, we'll need to write scenarios that will lead to step definitions that start the RESTful API service.

There's an important question about setting the state of the RESTful API server. One approach to setting a state is by making a sequence of requests as part of the scenario. This is often appropriate for this application.

If the server's state is reflected in the file system, then seeding proper files can be a way to control the state of the API server. Rather than run an acquire and clean process, a test scenario can inject the appropriate status and log files into a working directory.

Some developers have a feeling that a database (or a distributed cache) is required for RESTful APIs. In practice, it's often the case that a shared file system is sufficient.

 Using files is not uncommon in practice. A database to share state is not **always** required for RESTful APIs.

Using the file system for the state makes acceptance testing work out nicely. The proper files can be created to initialize the service in the state described by the given steps in the test scenario.

A complicated scenario could look like the following:

```
@fixture.REST_server
Scenario: Service starts and finishes acquiring data.

  Given initial request is made with path "/api/2023.02/creation" and
     method "post" and
```

```
        body with {"series": "2", "source": "Anscombe_quartet_data.csv"}
    And initial response has status "200", content-type "application/json"
    And initial response has job-id
    When polling every 2 seconds with path "/api/2023.02/creation/job-id" and
        method "get" finally has response body with status "Done"
    Then response content-type is "application/json"
    And response body has log with more than 0 lines
    And response body has series "Series_2"
    And response body has status "done"
```

For more background on creating a fixture, see *Acceptance tests* in *Chapter 4, Data Acquisition Features: Web APIs and Scraping*. This scenario references a fixture named REST_server. This means the environment.py must define this fixture, and provide a before_tag() function that will make sure the fixture is used.

The given steps specify an initial query and response. This should set the required state in the API server. This request for processing will initiate the acquire and clean processing. The When step specifies a sequence of actions that include polling periodically until the requested processing finishes.

Note the path provided in the When statement. The text job-id is in the scenario's path. The step definition function must replace this template string with the actual job identifier. This identifier will be in response to the initial request in the given step. The Given step's definition function must save the value in the context for use in later steps.

The Then step confirms that series data was returned. This example does not show a very complete check of the result. You are encouraged to expand on this kind of acceptance test scenario to be more complete in checking the actual results match the expected results.

For some applications, the retrieval of a tiny test case dataset may be a feature that helps test the application. The ordinary datasets the users want may be quite large, but a special, exceptionally small dataset may also be made available to confirm all the parts are working in concert.

A self-test resource is often essential for health checks, diagnostics, and general site reliability.

Network load balancers often need to probe a server to be sure it's capable of handling requests. A self-test URI can be helpful for this purpose.

A very subtle issue arises when trying to stop this service. It contains a worker pool, and the parent process needs to use the Linux wait() to properly terminate the children.

One reliable way to do this is to use server.send_signal(signal.SIGINT) in the function that starts the service to create the fixture for a scenario. This means the fixture function will have the following outline:

```
@fixture
def rest_server(context: Context) -> Iterator[Any]:
    # Create log file, base URI (code omitted)

    server = subprocess.Popen([sys.executable, "src/service.py"],
    shell=False, stdout=context.log_file, stderr=subprocess.STDOUT)
    time.sleep(0.5)  # 500 ms delay to allow the service to open a socket

    yield server  # Scenario can now proceed.

    # 100 ms delay to let server's workers become idle.
    time.sleep(0.10)
    server.send_signal(signal.SIGINT)
    # 100 ms delay to let API's subprocesses all terminate.
    time.sleep(0.10)
```

The various sleep() timings are generous over-estimations of the time required for the server subprocess to complete the various startup and shut-down tasks. In some cases, the OS scheduler will handle this gracefully. In other cases, however, disconnected child processes can be left in the list of running processes. These "zombie processes" need to be terminated manually, something we'd like to avoid.

 On most Linux-derived OSs, the `ps -ef` command will show all processes. The `ps -ef | grep python` pipeline will show all Python processes.

From this list, any zombie worker pool processes should be apparent.

`signal.SIGINT` is the control-C interrupt signal. The Python process makes this an exception that will not be handled. When this exception exits from the `with` statement that created the process pool, a complete clean-up will be finished and no zombie processes will be left running.

Now that we've looked at the acceptance test that defines proper behavior, we can look at the RESTful API server application.

RESTful API app

The RESTful API application can be built with any of the available frameworks. Since a previous chapter (*Chapter 4, Data Acquisition Features: Web APIs and Scraping*) used the Bottle framework, you can continue with this small framework. Because Bottle is very much like Flask, when additional features are needed, the upgrade to Flask isn't horribly complicated.

One of the advantages of using Flask for this application is an integrated client for writing unit test cases. The Bottle project can do everything that's required, but it lacks a test client. When looking at unit testing, we'll also look at unit test tools for the Bottle framework.

In *OpenAPI 3 specification* we looked at the OpenAPI specification for a specific path. Here's how that specification can be implemented:

```
from bottle import response, get


@get('/api/2023.02/series')
def series_list():
```

```
series_metadata = [
    {"name": series.stem, "elements": series_size(series)}
    for series in DATA_PATH.glob("*.ndj")
]
response.status = 200
response.body = json.dumps(series_metadata, indent=2)
response.content_type = "application/json"
return response
```

This function builds a sequence of metadata dictionaries. Each item has a series name, which is used in a separate request to get the data. The size is computed by a small function to read the series and find the number of samples.

The `response` object is not always manipulated as shown in this example. This is an extreme case, where the value to be returned is not a Python dictionary. If the return value is a dictionary, the Bottle framework will convert it to JSON, and the content type will be set to `application/json` automatically. In this case, the result is a list of dictionaries; the Bottle framework will not automatically serialize the object in JSON notation.

An important part of the design is a cache to retain `Future` objects until the processing completes, and the data is available. One way to handle this is with a dataclass that keeps the parameters of the request, the `Future` object that will produce the results, and the assigned job identifier.

This structure for each `Future` object might look like the following example:

```
from conccurrent import futures
from dataclasses import dataclass, field
from pathlib import Path
import secrets


@dataclass
class AcquireJob:
```

```
    series: str
    source_path: Path
    output_path: Path
    future: futures.Future = field(init=False)
    job_id: str = field(default_factory=lambda:
    \secrets.token_urlsafe(nbytes=12))
```

This keeps the parameters for the request as well as the processing details. The values for `series`, `source_path`, and `output_path` are built from the parameters provided when making an initial request. The paths are built from supplied names and include the base path for the working directory the server is using. In this example, the user's input is limited to the series name and the data source. These come from a small domain of valid values, making it relatively easy to validate these values.

The RESTful API can then create the output path within the appropriate directory of cleaned data.

The value for the `job_id` attribute is computed automatically when an instance of the `AcquireJob` class is created.

The value for the `future` attribute is set when the `submit()` method is used to submit a processing request to process a pool of waiting workers.

The worker pool needs to be created before any work can be done by the RESTful API. The startup can look like the following:

```python
from conccurrent import futures
import urllib.parse


WORKERS: futures.ProcessPoolExecutor


# Definitions of all of the routes


if __name__ == "__main__":
```

```
        # Defaults...
        acquire_uri = "http://localhost:8080"
        # Parse a configuration, here; possibly overriding defaults
    uri = urllib.parse.urlparse(acquire_uri)
        with futures.ProcessPoolExecutor() as WORKERS:
        run(host=uri.hostname, port=uri.port)
```

Each route is handled by a separate function. Because of this, the Bottle (as well as the Flask) framework expects the worker pool to be a global object shared by all of the route-handling functions. In the event of a multi-threaded server, a lock must be used before a write access to the WORKERS global.

Similarly, the cache of AcquireJob instances is also expected to be a global object. This is updated only by the route-handling function to handle initiating a processing request. This cache will be queried by a route that shows the status of a processing request. In the event of a multi-threaded server, a lock must be used before adding a new item to the global cache of working jobs.

In some cases, where the load is particularly heavy, thread-local storage may be needed for any processing done by the various functions in the RESTful API implementation. The request and response objects, in particular, are already in thread-local storage. Ideally, there is very little processing done by these functions, minimizing the number of objects that need to be created and kept in an instance of threading.local.

There are a few special considerations for the unit tests for this project. We'll look at those in the next section.

Unit test cases

Some frameworks — like **Flask** — offer a test client that can be used to exercise an application without the overheads of starting a server and a worker pool.

The **Bottle** framework doesn't offer a test client. An associated project, **boddle**, offers a way to build a mock request object to support unit testing. See https://github.com/ker

`edson/boddle`.

The **WebTest** project is an alternative for writing unit tests. A **WebTest** fixture contains the Bottle application and provides requests and responses through the internal WSGI interface. This avoids the need to start a complete server. It also permits some monkey-patching of the Bottle application to mock components. See `https://docs.pylonsproject.org/projects/webtest/en/latest/`.

It seems best to use the very sophisticated `WebTest` client that's part of the **Pylons** framework. This client can execute the unit tests.

It's sometimes helpful to note that functions with decorators are composite objects. This means the "unit" test isn't testing the decoration and the function in isolation from each other. This lack of separate testing can sometimes lead to difficulty in debugging the root cause of a test case failure. A problem may be in the function, it may be the `@route` decorator, or it may be any authorization decorator that's also part of the composite function being tested.

It seems easier to test the composite route functions, using appropriate log messages for debugging. While this doesn't follow the strict idea of testing each component in isolation, it does work well for testing each route with appropriate mocks. For example, we can mock the worker pool, avoiding the overhead of starting a subprocess when testing.

Here's an example of a test function using **WebTest** to exercise a **Bottle** route:

```
from unittest.mock import sentinel, Mock, call
from pytest import fixture, MonkeyPatch
from webtest import TestApp
import service


def test_test(monkeypatch: MonkeyPatch) -> None:
    monkeypatch.setitem(service.ACCESS, "unit-test", "unit-test")
    app = TestApp(service.app)
    app.authorization = (
```

```
        "Basic", ("unit-test", "unit-test")
    )
    response = app.get("/api/2023.02/test")
    assert response.status_code == 200
    assert response.json['status'] == "OK"
```

`service.app` is the global `app` object in the RESTful API application. This is an instance of the `Bottle` class. `service.ACCESS` is the global list of usernames and their expected API keys. This is monkey-patched by the test to force in a specific test username and test API Key. This initial setup is something that might be used by a number of tests and should be defined as a reusable fixture.

When the `app.get()` request is made, the test harness will execute the `route` function and collect the response for examination by the `test` method. This makes a direct function call, avoiding the overhead of a network request.

One of the reasons for choosing to use **Flask** instead of **Bottle** is the availability of a test client that can simplify some of this test setup.

Summary

This chapter integrated a number of application programs under the cover of a single RESTful API. To build a proper API, there were several important groups of skills:

- Creating an OpenAPI specification.

- Writing a web service application to implement the OpenAPI specification.

- Using a processing pool to delegate long-running background tasks. In this example, we used `concurrent.futures` to create a future promise of results, and then compute those results.

The number of processes involved can be quite daunting. In addition to the web service, there is a processing pool, with a number of sub-processes to do the work of acquiring and cleaning data.

In many cases, additional tools are built to monitor the API to be sure it's running properly. Further, it's also common to allocate dedicated servers to this work, and configure `supervisord` to start the overall service and ensure the service continues to run properly.

Extras

Here are some ideas for you to add to these projects.

Add filtering criteria to the POST request

The **POST** request that initiates acquire processing is quite complicated. See *A POST request starts processing* to see the processing it does.

We might name the function for this route `creation_job_post()` to make it clear that this creates jobs to acquire data in response to an HTTP POST request.

The list of tasks in this function includes the following:

1. Check the user's permissions.

2. Validate the parameters.

3. Build an `AcquireJob` instance with the parameters.

4. Update the `AcquireJob` instance with the `Future` object. The future will evaluate the `acquire_series()` function that does the work of acquiring and cleaning the data.

5. Return a JSON object with details of the submitted job, as well as headers and a status code to redirect to a request to get the job's status.

Some RESTful APIs will have even more complicated parameters. For example, users may want to filter the data to create a subset before downloading. This improves the UX by providing only the required data. It also allows analysts to share subsets of data without having to share the filtering code within the analyst community.

It can also improve the UX by performing filtering on larger, powerful servers. It can prevent having to download and filter data on a local laptop.

This is emphatically **not** a feature of the RESTful API. This must **first** be built as a feature of an application that reads and filters the clean data. This new application will create a new dataset, ready for download. The data set name might be a UUID, and an associated metadata file would contain the filter parameters.

The implementation requires the `creation_job_post()` function to now also validate the filter criteria. It must include the filter criteria in the `AcquireJob` instance that is built, and provide the filter criteria to the underlying `acquire_series()` function.

The `acquire_series()` function will have the most dramatic changes. It will run the acquire, clean, and filter applications as subprocesses. You may want to consider an integrated application that runs the other applications, simplifying the RESTful API.

This will, of course, lead to considerably more complicated acceptance test cases to be sure the data acquisition works with — and without — these additional filter criteria.

Split the OpenAPI specification into two parts to use $REF for the output schema

The OpenAPI specification includes a number of schema. In *OpenAPI 3 specification*, we showed a few key features of this specification.

It's not too difficult for an analyst to download the entire specification, and then locate the `components.schemas.seriesList` schema. This navigation through a JSON document doesn't involve too many challenges.

While this is not burdensome, some users might object. An analyst focused on a business problem should not be asked to also sort out the structure of the OpenAPI specification. An alternative is to decompose the specification into pieces and serve the pieces separately.

Specifically, the places where `"$ref"` references appear generally use a path of the form `#/components/schemas/....` The path is a local URL, omitting the hostname information. This can be replaced with a full URL that refers to schema details on the RESTful API server.

We might use `http://localhost:8080/api/schemas/...` to refer to the various schema files stored as separate JSON documents. Each individual schema definition would have a distinct URI, permitting ready access to only the relevant schema, and ignoring other aspects of the OpenAPI specification.

This decomposes the OpenAPI specification into the overall specification for the service and separate specifications for a schema that describes downloadable datasets. It also requires adding a path to the RESTful API service to properly download the schema in addition to downloading the overall OpenAPI specification.

This leads to a few extra acceptance test cases to extract the schema as well as the overall OpenAPI specification.

Use Celery instead of concurrent.futures

The suggestion in *Overall approach* is to use the `concurrent.futures` module to handle the long-running data acquisition and cleaning processes. The API requests that initiate processing create a `Future` object that reflects the state of a separate subprocess doing the actual work. The RESTful API is free to respond to additional requests while the work is being completed.

Another popular package for implementing this kind of background processing is `celery`. See `https://docs.celeryq.dev/en/stable/getting-started/introduction.html`.

This is a bit more complicated than using the `concurrent.futures` module. It also scales elegantly to allow a large number of separate computers to comprise the pool of available workers. This can permit very large processing loads to be controlled by a relatively small RESTful API application.

Using Celery requires creating tasks, using the `@task` decorator. It also requires starting the worker pool separately. This means the overall RESTful API now has two steps to get started:

- The celery worker pool must be running.

- The RESTful API can then start. Once it's running, it can delegate work to workers in the pool.

For very large workloads, where the worker pool is spread across multiple computers, use of Celery's sophisticated management tools are required to be sure the pools are starting and stopping appropriately.

The core work of submitting work to the worker pool changes from `pool.submit()` to `celery_app.delay()`. This is a small programming change that permits using a more sophisticated and scalable worker pool.

There aren't any acceptance test changes for this. The features are identical.

The fixture definition required to start the RESTful API will be more complicated: it will have to start the **Celery** pool of workers before starting the RESTful API. It will also need to shut down both services.

Call external processing directly instead of running a subprocess

In *Overall approach*, we suggested the work should be done by an `acquire_series()` function. This function would be evaluated by the `POOL.submit()` function. This would delegate the work to a worker, and return a `Future` object to track the state of completion.

In that section, we suggested the `acquire_series()` function could use `subprocess.run()` to execute the various components of the processing pipeline. It could run the `src/acquire.py` application, and then run the `src/clean.py` application, using the `subprocess` module.

This isn't the only way it could work. The alternative is to import these application modules, and evaluate their `main()` functions directly.

This means replacing the `subprocess.run()` function with the `acquire.main()` and `clean.main()` functions. This avoids a tiny overhead in Linux. It can be a conceptual simplification to see how the `acquire_series()` function creates the data using other Python modules.

This involves no changes to the acceptance test cases. It does involve some changes to the unit test cases. When using `subprocess.run()`, the unit test must monkey-patch the `subprocess` module with a mock that captures the argument values and returns a useful result. When replacing this processing with the `acquire.main()` and `clean.main()` functions, these two modules must be monkey patched with mocks that capture the argument values and return useful results.

13
Project 4.1: Visual Analysis Techniques

When doing **exploratory data analysis (EDA)**, one common practice is to use graphical techniques to help understand the nature of data distribution. The US **National Institute of Standards and Technology (NIST)** has an *Engineering Statistics Handbook* that strongly emphasizes the need for graphic techniques. See `https://doi.org/10.18434/M32189`.

This chapter will create some additional Jupyter notebooks to present a few techniques for displaying univariate and multivariate distributions.

In this chapter, we'll focus on some important skills for creating diagrams for the cleaned data:

- Additional Jupyter Notebook techniques

- Using **PyPlot** to present data

- Unit testing for Jupyter Notebook functions

This chapter has one project, to build the start of a more complete analysis notebook. A notebook can be saved and exported as a PDF file, allowing an analyst to share preliminary results for early conversations. In the next chapter, we'll expand on the notebook to create a presentation that can be shared with colleagues.

Looking further down the road, a notebook can help to identify important aspects of the data that need ongoing monitoring. The computations created here will often become the basis for more fully automated reporting tools and notifications. This analysis activity is an important step toward understanding the data and designing a model for the data.

We'll start with a description of an analysis notebook.

Description

In the previous chapters, the sequence of projects created a pipeline to acquire and then clean the raw data. The intent is to build automated data gathering as Python applications.

We noted that ad hoc data inspection is best done with a notebook, not an automated CLI tool. Similarly, creating command-line applications for analysis and presentation can be challenging. Analytical work seems to be essentially exploratory, making it helpful to have immediate feedback from looking at results.

Additionally, analytical work transforms raw data into information, and possibly even insight. Analytical results need to be shared to create significant value. A Jupyter notebook is an exploratory environment that can create readable, helpful presentations.

One of the first things to do with raw data is to create diagrams to illustrate the distribution of univariate data and the relationships among variables in multivariate data. We'll emphasize the following common kinds of diagrams:

- **Histograms** A histogram summarizes the distribution of values for a variable in a dataset. The histogram will have data values on one axis and frequency on the other axis.

- **Scatter Plots** A scatter plot summarizes the relationships between values for two variables in a dataset. The visual clustering can be apparent to the casual observer.

For small datasets, each relationship in a scatter plot can be a single dot. For larger datasets, where a number of points have similar relationships, it can be helpful to create "bins" that reflect how many points have the same relationship.

There are a number of ways of showing the size of these bins. This includes using a color code for more popular combinations. For some datasets, the size of an enclosing circle can show the relative concentration of similarly-valued data. The reader is encouraged to look at alternatives to help emphasize the interesting relationships among the attributes of the various samples.

The use of **Seaborn** to provide colorful styles is also important when working with diagrams. You are encouraged to explore various color palettes to help emphasize interesting data.

Overall approach

We'll take some guidance from the C4 model (`https://c4model.com`) when looking at our approach:

- **Context**: For this project, the context diagram has two use cases: the acquire-to-clean process and this analysis notebook.

- **Containers**: There's one container for analysis application: the user's personal computer.

- **Components**: The software components include the existing analysis models that provide handy definitions for the Python objects.

- **Code**: The code is scattered in two places: supporting modules as well as the notebook itself.

A context diagram for this application is shown in *Figure 13.1*.

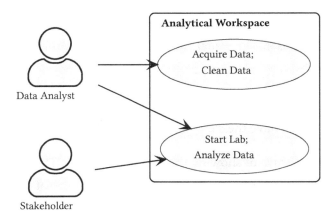

Figure 13.1: Context diagram

The analyst will often need to share their analytical results with stakeholders. An initial notebook might provide confirmation that some data does not conform to the null hypothesis, suggesting an interesting relationship that deserves deeper exploration. This could be part of justifying a budget allocation to do more analysis based on preliminary results. Another possible scenario is sharing a notebook to confirm the null hypothesis is likely true, and that the variations in the data have a high probability of being some kind of measurement noise. This could be used to end one investigation and focus on alternatives.

In *Chapter 6, Project 2.1: Data Inspection Notebook*, the data inspection notebook was described. The use of a single acquisition model module for the data was mentioned, but the details of the Python implementation weren't emphasized. The raw data module often provides little useful structure to the data when doing inspections.

Moving forward into more complicated projects, we'll see the relationship between the notebooks and the modules that define the data model become more important.

For this analysis notebook, the analysis data model, created in *Chapter 9, Project 3.1: Data Cleaning Base Application*, will be a central part of the notebook for doing analysis. This will be imported and used in the analysis notebook. The analysis process may lead to changes to the analysis model to reflect lessons learned.

A technical complication arises from the directory structure. The data acquisition and cleaning applications are in the `src` directory, where the notebooks are kept in a separate `notebooks` directory.

When working with a notebook in the `notebooks` directory, it is difficult to make the Python `import` statement look into the adjacent `src` directory. The `import` statement scans a list of directories defined by the `sys.path` value. This value is seeded from some defined rules, the current working directory, and the value of the `PYTHONPATH` environment variable.

There are two ways to make the `import` statement load modules from the adjacent `src` directory:

1. Put `../src` into the `PYTHONPATH` environment variable before starting JupyterLab.

2. Put the absolute path to `../src` into the `sys.path` list after starting JupyterLab.

The two are equivalent. The first can be done by updating one's `~/.zshrc` file to make sure the `PYTHONPATH` environment variable is set each time a terminal session starts. There are other files appropriate for other shells; for example, `~/.bashrc` or `./rc` for the classic `sh` shell. For Windows, there's a dialog that allows one to edit the system's environment variables.

The alternative is to update `sys.path` with a cell containing code like the following example:

```python
import sys
from pathlib import Path


src_path = Path.cwd().parent / "src"
sys.path.append(str(src_path))
```

This cell will add the peer `../src` directory to the system path. After this is done, the `import` statement will bring in modules from the `../src` directory as well as bringing in built-in standard library modules, and modules installed with **conda** or **pip**.

An initialization module can be defined as part of the IPython startup. This module can alter the `sys.path` value in a consistent way for a number of related notebooks in a project.

While some developers object to tinkering with sys.path in a notebook, it has the advantage of being explicit.

Setting PYTHONPATH in one's ~/.zshrc file is a very clean and reliable solution. It then becomes necessary to put a reminder in a README file so that new team members can also make this change to their personal home directory.

When sharing notebooks, it becomes imperative to make sure all stakeholders have access to the entire project the notebook depends on. This can lead to a need to create a Git repository that contains the notebook being shared along with reminders, test cases, and needed modules.

Once we have the path defined properly, the notebook can share classes and functions with the rest of the applications. We'll move on to looking at one possible organization of an analysis notebook.

General notebook organization

An analytical notebook is often the primary method for presenting and sharing results. It differs from a "lab" notebook. It's common for lab notebooks to contain a number of experiments, some of which are failures, and some of which have more useful results. Unlike a lab notebook, an analytical notebook needs to have a tidy organization with carefully-written Markdown cells interspersed with data processing cells. In effect, an analysis notebook needs to tell a kind of story. It needs to expose actors and actions and the consequences of those actions.

It's essential that the notebook's cells execute correctly from beginning to end. The author should be able to restart the kernel and run all of the cells at any time to redo the computations. While this may be undesirable for a particularly long-running computation, it still must be true.

A notebook may have preliminary operations that aren't relevant to most readers of the report. A specific example is setting the sys.path to import modules from an adjacent

../src directory. It can be helpful to make use of JupyterLab's ability to collapse a cell as a way to set some of the computation details aside to help the reader focus on the key concepts.

It seems sensible to formalize this with Markdown cells to explain the preliminaries. The remaining cells in this section can be collapsed visually to minimize distractions.

The preliminaries can include technical and somewhat less technical aspects. For example, setting sys.path is purely technical, and few stakeholders will need to see this. On the other hand, reading the SeriesSample objects from an ND JSON format file is a preliminary step that's somewhat more relevant to the stakeholder's problems.

After the preliminaries, the bulk of the notebook can focus on two topics:

- Summary statistics

- Visualizations

We'll look at the summary statistics in the next section.

Python modules for summarizing

For initial reports, Python's statistics module offers a few handy statistical functions. This module offers mean(), median(), mode(), stdev(), variance(). There are numerous other functions here the reader is encouraged to explore.

These functions can be evaluated in a cell, and the results will be displayed below the cell. In many cases, this is all that's required.

In a few cases, though, results need to be truncated to remove meaningless trailing digits. In other cases, it can help to use an f-string to provide a label for a result. A cell might look like the following:

```
f"mean = {statistics.mean(data_values): .2f}"
```

This provides a label, and truncates the output to two places to the right of the decimal point.

In some cases, we might want to incorporate computed values into markdown cells. The **Python Markdown** extension provides a tidy way to incorporate computed values into markdown content.

See `https://jupyter-contrib-nbextensions.readthedocs.io/en/latest/nbextensions/python-markdown/readme.html`.

The most important part of the notebook is the graphic visualization of the data, we'll turn to that in the next section.

PyPlot graphics

The **matplotlib** package works well for creating images and diagrams. Within this extensive, sophisticated graphic library is a smaller library, `pyplot`, that's narrowly focused on data visualizations. Using the `pyplot` library permits a few lines of code to create a useful display of data.

The module is often renamed to make it easier to type. The common convention is shown in the following line of code:

```
import matplotlib.pyplot as plt
```

This lets us use `plt` as a namespace to refer to functions defined in the `pyplot` module of the `matplotlib` package.

In some cases, JupyterLab may not have the `matplotlib` library prepared for interactive use. (This will be clear when the plotted images are not shown in the notebook.) In these cases, the interactive use of the `matplotlib` library needs to be enabled. Use the following magic commands in a cell of the notebook to enable interactive use:

```
%matplotlib inline
%config InlineBackend.figure_formats = {'png', 'retina'}
```

The first command enables the `matplotlib` library to create graphics immediately in the notebook. It won't create separate files or pop-up windows. The second command produces readily-shared PNG output files. It also helps **MAC OS X** users to optimize the graphics for their high-resolution displays.

This isn't always required. It's needed only for those installations that didn't configure the `matplotlib` library for interactive use in a Jupyter notebook.

In many cases, the `pyplot` library expects simple sequences of values for the various plotting functions. The *x* and *y* values of a scatter plot, for example, are expected to be two parallel lists of the same length. This will lead to a few extra cells to restructure data.

The data cleaning applications from previous chapters produced a single sequence of compound sample objects. Each object was a separate sample, with the related values for all of the variables. We'll need to convert from this sample record organization to parallel sequences of values for the individual variables.

This can involve something like the following:

```
x = [s.x for s in series_1]
y = [s.y for s in series_1]
```

An alternative is to use the `operator.attrgetter()` function. It looks like the following:

```
from operator import attrgetter


x = list(map(attrgetter('x'), series_1))
y = list(map(attrgetter('y'), series_1))
```

We'll start with a histogram to show the distribution of values for a single variable.

Data frequency histograms

Often, a document — like a book — has individual figures. A figure may contain a single plot. Or, it may contain a number of "subplots" within a single figure. The `pyplot` library provides support for creating a single figure that contains many subplots. The idea of a figure with a single plot can be seen as a special case of this generalized approach.

A figure with a single plot can be prepared with a statement like the following:

```
fig, ax = plt.subplots()
```

The `fig` object is the figure as a whole. The `ax` object is the set of axes for the subplot within the figure. While the explicit `ax` object can seem unnecessary, it's part of a more general approach that allows figures with related plots to be built.

The more general case stacks multiple subplots within a figure. The `sublots()` function can return axes for each subplot. The call to create a two-plot figure might look like this:

```
fig, (ax_0, ax_1) = plt.subplots(1, 2)
```

Here we've stated we'll have 1 row with 2 subplots stacked next to each other. The `fig` object is the figure as a whole. The `ax_0` and `ax_1` objects are the sets of axes for each of these two subplots within the figure.

Here's an example of creating a single histogram from the y values of one of the four Anscombe's quartet series:

```
fig, ax = plt.subplots()


# Labels and Title
ax.set_xlabel('Y')
ax.set_ylabel('Counts')
ax.set_title('Series I')


# Draw Histogram
_ = ax.hist(y, fill=False)
```

For this example, the y variable must be a list of *y* attribute values from Series I. The series data must be read from the cleaned source file.

The `_ = ax.hist(...)` statement assigns the results of the `hist()` function to the variable `_` as a way to suppress displaying the result value in this cell. Without this assignment statement, the notebook will show the result of the `hist()` function, which isn't very interesting and clutters up the output. Since each series has both an *x* and a *y* value, it helps to stack two histograms that shows the distribution of these values. The reader is encouraged to develop a figure with two stacked subplots.

The number of options for color and borders on the histogram bars are breathtaking in their complexity. It helps to try several variants in a single notebook and then delete the ones that aren't helpful.

The comparison between values is often shown in a scatter plot that shows each (x, y) pair. We'll look at this next.

X-Y scatter plot

A scatter plot is one of many ways to show the relationship between two variables. Here's an example of creating a single plot from the x and y values of one of the four Anscombe's quartet series:

```
fig, ax = plt.subplots()


# Labels and Title
ax.set_xlabel('X')
ax.set_ylabel('Y')
ax.set_title('Series I')


# Draw Scatter
_ = ax.scatter(x, y)
```

The _ = ax.scatter(...) statement assigns the results of the scatter() function to the variable _ as a way to suppress displaying the value. This keeps the output focused on the figure.

The above example presumes the data have been extracted from a sequence of SeriesSample instances using code like the following:

```
x = [s.x for s in series_1]
y = [s.y for s in series_1]
```

This, in turn, presumes the value for series_1 was read from the clean data created by the

acquire and **clean** applications.

Now that we have an approach to building the notebook, we can address the way notebooks tend to evolve.

Iteration and evolution

A notebook is often built iteratively. Cells are added, removed, and modified as the data is understood.

On a purely technical level, the Python programming in the cells needs to evolve from a good idea to software that's testable. It's often easiest to do this by rewriting selected cells into functions. For example, define a function to acquire the ND JSON data. This can be supplemented by `doctest` comments to confirm the function works as expected.

A collection of functions can be refactored into a separate, testable module if needed. This can permit wider reuse of good ideas in multiple notebooks.

Equally important is to avoid "over-engineering" a notebook. It's rarely worth the time and effort to carefully specify the contents of a notebook, and then write code that meets those specifications. It's far easier to create — and refine — the notebook.

In some large organizations, a senior analyst may direct the efforts of junior analysts. In this kind of enterprise, it can be helpful to provide guidance to junior analysts. When formal methods are needed, the design guidance can take the form of a notebook with markdown cells to explain the desired goal.

Now that we have an approach to building the notebook, we can enumerate the deliverables for this project.

Deliverables

This project has the following deliverables:

- A `requirements-dev.txt` file that identifies the tools used, usually `jupyterlab==3.5.3` and `matplotlib==3.7.0`.

- Documentation in the docs folder.

- Unit tests for any new application modules in the tests folder.

- Any new application modules in the src folder with code to be used by the inspection notebook.

- A notebook to summarize the clean data. In the case of Anscombe's quartet, it's essential to show the means and variances are nearly identical, but the scatter plots are dramatically different.

We'll look at a few of these deliverables in a little more detail.

Unit test

There are two distinct kinds of modules that can require testing:

- The notebook with any function or class definitions. All of these definitions require unit tests.

- If functions are factored from the notebook into a supporting module, this module will need unit tests. Many previous projects have emphasized these tests.

A notebook cell with a computation cell is notoriously difficult to test. The visual output from the hist() or scatter() functions seems almost impossible to test in a meaningful way.

In addition, there are numerous usability tests that can't be automated. Poor choice of colors, for example, can obscure an important relationship. Consider the following questions:

1. Is it informative?

2. Is it relevant?

3. Is it misleading in any way?

In many cases, these questions are difficult to quantify and difficult to test. As a consequence, it's best to focus automated testing on the Python programming.

It's imperative to avoid testing the internals of `matplotlib.pyplot`.

What's left to test?

- The data loading.

- Any ad hoc transformations that are part of the notebook.

The data loading should be reduced to a single function that creates a sequence of `SeriesSample` instances from the lines in an ND JSON file of clean data. This loading function can include a test case.

We might define the function as follows:

```python
def load(source_file):
    """
    >>> from io import StringIO
    >>> file = StringIO('''{"x": 2.0, "y": 3.0}\\n{"x": 5.0, "y": 7.0}''')
    >>> d = load(file)
    >>> len(d)
    2
    >>> d[0]
    SeriesSample(x=2.0, y=3.0)
    """
    data = [
        SeriesSample(**json.loads(line)) for line in source_file if line
    ]
    return data
```

This permits testing by adding a cell to the notebook that includes the following:

```python
import doctest
doctest.testmod()
```

This cell will find the functions defined by the notebook, extract any doctest cases from the docstrings in the function definitions, and confirm the doctest cases pass.

For more complicated numerical processing, the **hypothesis** library is helpful. See *Hypothesis testing* in *Chapter 10, Data Cleaning Features* for more information.

Acceptance test

An automated acceptance test is difficult to define for a notebook. It's hard to specify ways in which a notebook is helpful, meaningful, or insightful in the simple language of Gherkin scenarios.

The `jupyter execute <filename>` command will execute an `.ipynb` notebook file. This execution is entirely automated, allowing a kind of sanity check to be sure the notebook runs to completion. If there is a problem, the command will exit with a return code of 1, and the cell with the error will be displayed in detail. This can be handy for confirming the notebook isn't trivially broken.

The `.ipynb` file is a JSON document. An application (or a step definition for **Behave**) can read the file to confirm some of its properties. An acceptance test case might look for error messages, for example, to see if the notebook failed to work properly.

Cells with `"type": "code"` will also have `"outputs"`. If one of the outputs has `"output_type": "error"`; this cell indicates a problem in the notebook. The notebook did not run to completion, and the acceptance test should be counted as a failure.

We can use projects like **Papermill** to automate notebook refresh with new data. This project can execute a template notebook and save the results as a finalized output notebook with values available and computations performed.

For more information, see `https://papermill.readthedocs.io`.

Summary

This project begins the deeper analysis work on clean data. It emphasizes several key skills, including:

- More advanced Jupyter Notebook techniques. This includes setting the `PYTHONPATH` to import modules and creating figures with plots to visualize data.

- Using **PyPlot** to present data. The project uses popular types of visualizations: histograms and scatter plots.

- Unit testing for Jupyter Notebook functions.

In the next chapter, we'll formalize the notebook into a presentation "slide deck" that can be shown to a group of stakeholders.

Extras

Here are some ideas for the reader to add to these projects.

Use Seaborn for plotting

An alternative to the **pyplot** package is the **Seaborn** package. This package also provides statistical plotting functions. It provides a wider variety of styling options, permitting more colorful (and perhaps more informative) plots.

See `https://seaborn.pydata.org` for more information.

This module is based on `matplotlib`, making it compatible with JupyterLab.

Note that the **Seaborn** package can work directly with a list-of-dictionary structure. This matches the ND JSON format used for acquiring and cleaning the data.

Using a list-of-dictionary type suggests it might be better to avoid the analysis model structure, and stick with dictionaries created by the **clean** application. Doing this might sacrifice some model-specific processing and validation functionality.

On the other hand, the `pydantic` package offers a built-in `dict()` method that covers a sophisticated analysis model object into a single dictionary, amenable to use with the **Seaborn** package. This seems to be an excellent way to combine these packages. We encourage the reader to explore this technology stack.

Adjust color palettes to emphasize key points about the data

Both the **pyplot** package and the **Seaborn** package have extensive capabilities for applying color to the plot. A choice of colors can sometimes help make a distinction visible, or it can obscure important details.

You can consider alternative styles to see which seems more useful.

In some enterprise contexts, there are enterprise communications standards, with colors and fonts that are widely used. An important technique when using Seaborn is to create a style that matches enterprise communication standards.

A number of websites provide website color schemes and design help. Sites like `https://paletton.com` or `colordesigner.io` provide complementary color palettes. With some effort, we can take these kinds of designs for color palettes and create Seaborn styles that permit a consistent and unique presentation style.

14

Project 4.2: Creating Reports

One easy way to share good-looking results is to use a Jupyter notebook's **Markdown** cells to create a presentation. This chapter will create a "slide deck" that can be shared and presented. We can expand on this to create a PDF report using additional packages like **Jupyter book** or **Quarto**.

In this chapter, we'll look at two important working results of data analysis:

- Slide decks build directly from a Jupyter Lab notebook.

- PDF reports built from notebook data and analysis.

This chapter's project will upgrade an analysis notebook created in the previous chapter to create presentations that can be shared with colleagues. We'll start by looking at the kinds of reports an analyst may need to produce.

Description

The first dozen chapters in the book created a pipeline to acquire and clean raw data. Once the data is available, we can now do more analytical work on the clean data.

The goal is to transform raw data into information, and possibly even insight to help stakeholders make properly informed decisions. Analytical results need to be shared to be valuable. A Jupyter Notebook is a solid basis to create readable, helpful presentations and reports.

We'll start by transforming an analysis notebook into a slide deck. You can then use this slide deck to talk through our key points with stakeholders, providing helpful visuals to back up the information they need to understand. These are common in an enterprise environment. (Some would argue they are too common and contain too much of the wrong kind of details.)

We'll start by looking at creating slide decks and presentations in Jupyter Lab.

Slide decks and presentations

A Jupyter Notebook can be exported to a presentation file. The underlying presentation will be an HTML-based repository of individual pages. **Reveal.js** project is used to control the navigation between pages. See `https://revealjs.com` for more details on how this engine works.

Within the notebook, each cell has properties. The "right sidebar" is the property inspector window, letting us manage these properties. One of the properties is the slide type. This allows the analyst to mark cells for inclusion in a presentation. We'll look at the technical details in *Preparing slides.*

There are a huge number of guides and tutorials on creating useful, informative presentations. The author likes to focus on three key points:

1. Tell them what you'll tell them. Present a list of topics (or an agenda or an outline). This will often use **Markdown** lists.

2. Tell them. This should proceed from general observations to the specific details of the presentation. This may be a mixture of **Markdown** text and figures.

3. Tell them what you told them. Present a summary of what your message was and the actions you'd like them to take in response. This, too, will often use **Markdown** lists.

What's important here is using the slide presentation to contain the keywords and phrases to help the audience remember the essential points and the call to action. This often means making use of **Markdown** text to emphasize words with a bold or italic font. It can also mean using **Markdown** lists of various kinds.

Another important part is avoiding visual clutter created when trying to cram too many points into a single page. When there are a lot of details, a presentation may not be the best approach to managing all the information. A report document may be more useful than a presentation. A document can provide supplemental details to support a brief presentation as well.

Reports

There's a blurry edge between a presentation and a report. Presentations tend to be shorter and focus on keywords and phrases. Reports tend to be longer and written in complete sentences. A well-organized presentation with complete sentences can be viewed as a brief report. A report with short paragraphs and a lot of figures can look like a presentation.

Markdown formatting provides a lot of capabilities to create publication-quality typeset documents. The technology stack from **Markdown** to HTML to a browser or a PDF rendering engine involves a number of transformation steps to get from simple Unicode text in a notebook cell to a richly detailed rendering. This stack is a first-class part of Jupyter Lab and can be exploited by tools like **Quarto** or **Jupyter{Book}** to create reports.

Not all of the typesetting conventions used by publications are available through **Markdown** source files. For example, some publication style guides will include an abstract section that has narrower margins, and sometimes a smaller font. This can be challenging to implement in the **Markdown** language. Some authors will use a less complex layout that lacks all the visual cues of margin and font size.

The power of HTML and CSS is such that a great many typesetting capabilities are available to the author willing to master the technology stack. The reader is encouraged to explore the capabilities of **Markdown**, HTML, and CSS. The reader is also advised to set realistic goals; a great deal of time can be invested in combining **Markdown** and CSS to achieve typesetting effects that don't enhance a report's message.

It often works out well to put each paragraph into a separate cell. This is not a strict rule: sometimes a group of paragraphs should be put into a single cell.

Top-level headings should often be in cells by themselves. This can make it easier to reorganize content within those headings. Some lower-level headings should be in the cell with their introductory paragraph since the heading and the cell's content are unlikely to be separated.

We might have a cell with a level one heading that looks like this:

```
# Title of our document
```

This cell has only the title and no following text. There will likely be a subheading in the next cell with the introduction to the document.

A lower-level cell might look like this:

```
## Conclusion
```

```
The various **Anscombe Quartet** series all have consistent
means and standard deviations.
```

This cell has both a level two title, and the introductory text for this section of the document. It uses ** **Markdown** syntax to show that a particular phrase should have strong emphasis, usually done with a bold font.

In the next sections, we'll talk about the technical approach to adding tools to the Jupyter environment so that the analyst can create presentations or reports.

Overall approach

We'll talk about the general technical steps to creating presentations and reports in a Jupyter Notebook. For presentations, no additional tools are needed. For some simple reports, the **File** menu offers the ability to save and export a notebook as pure Markdown, as a PDF file, or as a LaTeX document. For more complicated reports, it can help to use supplemental tools that create a more polished final document.

Preparing slides

An HTML-based presentation via **Reveal.js** is a first-class feature of a Jupyter Notebook. The **File** menu offers the ability to save and export a notebook as Reveal.js slides. This will create an HTML file that will display as a presentation.

Within Jupyter, the property inspector is used to set the type of slide for a cell. There's an icon of two meshed gears on the top right side of the page to show the property inspector in the right sidebar. Under the **View** menu, the option to show the right sidebar will also show the property inspector.

There are several choices of **Slide Type** for each cell. The most important two choices are "slide" and "skip".

The "slide" will be displayed as part of the presentation. The "skip" cells will be dropped from the presentation; this is great for computations and data preparation. The other options allow combining cells into a single slide and having subsidiary presentation slides.

Creating **Markdown** content and setting the slide type to "slide" creates the narrative text portion of a presentation. These slides would include title pages, agenda, and key points: all of the prompts and takeaway bullet points will be in these kinds of cells.

For data visualizations, we can use **Seaborn** or **PyPlot** to create the figure. The cell output has the slide type set to "slide" in the property inspector to include the visualization.

We can mark the computations, function definitions, and doctest cells with a slide type of **skip**. This will omit these details from the presentation.

The analyst can share the notebook with audience members who want to see the supporting details.

The **Reveal.js** has a huge repertoire of capabilities. Many of these features are available through HTML markup. For example, the auto-animate feature will smoothly transition between cells. Since HTML markup is part of **Markdown**, some familiarity with HTML is required for the use of the most advanced features.

The final step is to use the CLI to convert the notebook to a slide deck. The **File** menu has a **Save and Export Notebook As...** option, but this tends to make all of the code visible. Having visible code can distract from the essential message of the visualizations.

The following command hides the cell input value — the code — from the presentation:

```
jupyter nbconvert --to slides --no-input <notebook.ipynb>
```

Use the name of your notebook in place of <notebook.ipynb>. This will create an HTML file with the **Reveal.js** code included.

The overall process has three steps:

1. Edit the notebook.

2. Prepare the presentation. (The terminal tool in Jupyter Lab is ideal for this.)

3. View the presentation to find problems.

This is distinct from the way products like **Keynote** and **PowerPoint** work. When working with Jupyter Lab, there will a bit of flipping back- and forth between browser windows and the notebook window. Placing windows on each side of the display can be helpful.

Be sure to refresh the browser window after each change to the notebook.

Preparing a report

Creating reports is a first-class part of Jupyter Lab. The **File** menu offers the ability to save and export a notebook as pure Markdown, as a PDF file, or as a LaTeX document.

A tool like **pandoc** can convert a **Markdown** file into a wide variety of desired formats. For output creating using LaTeX formatting, a TeX rendering package is required to create a PDF file from the source. The **TeXLive** project maintains a number of tools useful for rendering LaTeX. For macOS users, the **MacTex** project offers the required binaries. An online tool like **Overleaf** is also useful for handling LaTeX.

In many cases, more sophisticated processing is required than simply saving a notebook as a pure **Markdown** file. We can add the **Jupyter{Book}** tools to our environment. For more information see `https://jupyterbook.org`.

The `jupyter-book` component needs to be added to the `requirements-dev.txt` file so other developers know to install it.

When using **conda** to manage virtual environments, the command might look like the following:

```
% conda install --channel conda-forge jupyter-book
```

When using other tools to manage virtual environments, the command might look like the following:

```
% python -m pip install jupyter-book
```

A Jupyter book can be considerably more complicated than a single notebook file. There will be configuration and **table-of-contents** (**TOC**) files to structure the overall report. The content can be provided in a mixture of **Markdown**, reStructuredText, and Notebook files. Additionally, an extended version of the **Markdown** language, **MyST**, is available to add a wide variety of semantic markup capabilities.

One way to get started is to use the `jupyter-book create` command to create a template project. This template includes the required `_config.yml` and `_toc.yml` files. It also includes examples of various other files that might be part of a project.

The `_config.yml` file has the title and author information. This is the place to start customizing the content to provide the correct report and author names. Other parts of the configuration may need to be changed, depending on how the report will be published. The built-in assumption is an HTML upload to a public repository. For many reports, this is ideal.

For some enterprise projects, however, reporting to a public repository with links to public GitHub isn't acceptable. For these cases, the `_config.yml` file will have to be changed to correct the repository options to refer to an in-house repository.

It's often helpful to immediately edit the `_toc.yml` file, and start creating the report's outline. Generally, the data and notebooks already exist. The audience is often known, and the key points the audience members need to absorb are clear, permitting the analyst to create the outline, and placeholder documents, right away.

In some cases, the analyst can fill in the outline with notes extracted from the analysis notebooks. This refactoring of content can help to trim working notebooks down to the essential computation and visualization. The narrative text can be segregated into MyST or Markdown files outside the notebooks.

Once the content is in draft form, a book is created with the `jupyter-book build` command. This will use the configuration and TOC file to build the complete document from various sources. The default document is an HTML page.

As of this book's publication date, version 0.15.1 includes a warning that direct PDF production is under development and may have bugs. The more reliable way to create PDFs is to use **Jupyter{Book}** to create a LaTeX file. The OS-native `LaTeX` command can be used to build PDFs. An alternative is to use the `sphinx-jupyterbook-latex` package to wrap the TeXtools that transform the LaTeX to PDF.

This involves a number of moving parts, and the installation can be daunting. Here are some of the steps involved:

1. The source **Markdown** text is converted to LATEX by Jupyter Book.

2. Some intermediate work may be performed by the `sphinx-jupyterbook-latex` package.

3. The final PDF is created by an OS `latex` command; this is either the **MacTex** or **TexLive** installation of the TEX tools.

The CLI build command is the `jupyter-book build` command with an additional option `--builder pdflatex` to specify that Sphinx and the TEX tools are used to render the PDF.

Creating technical diagrams

Technical diagrams, including the wide variety of diagrams defined by the UML, are often challenging to create. Popular presentation tools like **Keynote** and **PowerPoint** have clever drawing tools with lots of built-in shapes and options for positioning those shapes on a slide.

There are several choices for creating diagrams for a presentation:

- Use a separate graphics application to create `.PNG` or `.SVG` files and incorporate the graphics into the document. Many of the diagrams in this book were created with PlantUML, for example. See `https://plantuml.com`.

- Use `matplotlib`, and write code to create the image. This can involve a lot of programming to draw some boxes connected by arrows.

The **PlantWEB** project provides a Python interface to the PlantUML web service. This allows an analyst to work as follows:

1. Create a file with text in the **domain-specific language (DSL)** that describes the image.

2. Render the image with the PlantUML engine to create an `.SVG` file.

3. Import the image into the notebook as an IPython SVG object.

The image rendering uses the PlantUML server; this requires an active internet connection. In cases where the analyst might be working offline, the **PlantWEB** documentation suggests using **Docker** to run a local service in a local Docker container. This will do diagram rendering quickly without the requirement to connect to the internet.

Having looked at the various technical considerations to create slides and a report, we can emphasize the deliverables for this project.

Deliverables

There are two deliverables for this project:

- A presentation

- A report

The presentation should be an HTML document using the **Reveal.js** slide deck.

The report should be a PDF document from a single notebook. It should contain the visualization figures and some narrative text explaining the images.

For information on unit testing and acceptance testing of the notebooks, see *Chapter 13, Project 4.1: Visual Analysis Techniques*. This project should build on the previous project. It doesn't involve dramatic new programming. Instead, it involves the integration of a large number of components to create meaningful, useful presentations and reports.

Summary

In this chapter, we have built two important working results of data analysis:

- Slide decks that can be used as presentations to interested users and stakeholders

- Reports in PDF format that can be distributed to stakeholders

The line between these two is always hazy. Some presentations have a lot of details and are essentially reports presented in small pages.

Some reports are filled with figures and bullet points; they often seem to be presentations written in portrait mode.

Generally, presentations don't have the depth of detail reports do. Often, reports are designed for long-term retention and provide background, as well as a bibliography to help readers fill in missing knowledge. Both are first-class parts of a Jupyter notebook and creating these should be part of every analyst's skills.

This chapter has emphasized the additional tools required to create outstanding results. In the next chapter, we'll shift gears and look at some of the statistical basics of data modeling.

Extras

Here are some ideas for the reader to add to these projects.

Written reports with UML diagrams

In *Creating technical diagrams* the process of creating UML diagrams was summarized. The reader is encouraged to use PlantUML to create C4 diagrams for their data acquisition and cleaning pipeline. These `.SVG` files can then be incorporated into a report as Markdown figures.

For more information on the C4 model, see `https://c4model.com`.

15

Project 5.1: Modeling Base Application

The next step in the pipeline from acquisition to clean-and-convert is the analysis and some preliminary modeling of the data. This may lead us to use the data for a more complex model or perhaps machine learning. This chapter will guide you through creating another application in the three-stage pipeline to acquire, clean, and model a collection of data. This first project will create the application with placeholders for more detailed and application-specific modeling components. This makes it easier to insert small statistical models that can be replaced with more elaborate processing if needed.

In this chapter, we'll look at two parts of data analysis:

- CLI architecture and how to design a more complex pipeline of processes for gathering and analyzing data

- The core concepts of creating a statistical model of the data

Viewed from a distance, all analytical work can be considered to be creating a simplified model of important features of some complicated processes. Even something as simple-sounding as computing an average suggests a simple model of the central tendency of a variable. Adding a standard deviation suggests an expected range for the variable's values and – further – assigns a probability to values outside the range.

Models, can, of course, be considerably more detailed. Our purpose is to start down the path of modeling in a way that builds flexible, extensible application software. Each application will have unique modeling requirements, depending on the nature of the data, and the nature of the questions being asked about the data. For some processes, means and standard deviations are adequate for spotting outliers. For other processes, a richer and more detailed simulation may be required to estimate the expected distribution of data.

We'll start the modeling by looking at variables in isolation, sometimes called *univariate* statistics. This will examine a variety of commonly recognized distributions of data. These distributions generally have a few parameters that can be discovered from the given data. In this chapter, we'll also look at measures like mean, median, standard deviation, variance, and standard deviation. These can be used to describe data that has a normal or Gaussian distribution. The objective is to create create a CLI application separate from an analytic notebook used to present results. This creates a higher degree of automation for modeling. The results may then be presented in an analytical notebook.

There is a longer-term aspect to having automated model creation. Once a data model has been created, an analyst can also look at changes to a model and what implications the changes should have on the way an enterprise operates. For example, an application may perform a monthly test to be sure new data matches the established mean, median, and standard deviation reflecting the expected normal distribution of data. In the event that a batch of data doesn't fit the established model, further investigation is required to uncover the root cause of this change.

Description

This application will create a report on a dataset presenting a number of statistics. This automates the ongoing monitoring aspect of an Analysis Notebook, reducing the manual steps and creating reproducible results. The automated computations stem from having a statistical model for the data, often created in an analysis notebook, where alternative models are explored. This reflects variables with values in an expected range.

For industrial monitoring, this is part of an activity called **Gage repeatability and reproducibility**. The activity seeks to confirm that measurements are repeatable and reproducible. This is described as looking at a "measurement instrument." While we often think of an instrument as being a machine or a device, the definition is actually very broad. A survey or questionnaire is a measurement instrument focused on people's responses to questions.

When these computed statistics deviate from expectations, it suggests something has changed, and the analyst can use these unexpected values to investigate the root cause of the deviation. Perhaps some enterprise process has changed, leading to shifts in some metrics. Or, perhaps some enterprise software has been upgraded, leading to changes to the source data or encodings used to create the clean data. More complex still, it may be that the instrument doesn't actually measure what we thought it measured; this new discrepancy may expose a gap in our understanding.

The repeatability of the model's measurements is central to the usability of the measurements. Consider a ruler that's so worn down over years of use that it is no longer square or accurate. This single instrument will produce different results depending on what part of the worn end is used to make the measurement. This kind of measurement variability may obscure the variability in manufacturing a part. Understanding the causes of changes is challenging and can require thinking "outside the box" — challenging assumptions about the real-world process, the measurements of the process, and the model of those measurements.

Exploratory data analysis can be challenging and exhilarating precisely because there aren't obvious, simple answers to explain why a measurement has changed.

The implementation of this preliminary model is through an application, separate from the previous stages in the pipeline to acquire and clean the data. With some careful design, this stage can be combined with those previous stages, creating a combined sequence of operations to acquire, clean, and create the summary statistics.

This application will overlap with the analysis notebook and the initial inspection notebook. Some of the observations made during those earlier ad-hoc analysis stages will be turned into fixed, automated processing.

This is the beginning of creating a more complicated machine-learning model of the data. In some cases, a statistical model using linear or logistic regression is adequate, and a more complex artificial intelligence model isn't needed. In other cases, the inability to create a simple statistical model can point toward a need to create and tune the hyperparameters of a more complicated model.

The objective of this application is to save a statistical summary report that can be aggregated with and compared to other summary reports. The ideal structure will be a document in an easy-to-parse notation. JSON is suggested, but other easier-to-read formats like TOML are also sensible.

There are three key questions about data distribution:

1. What is the **location** or expected value for the output being measured?

2. What is the **spread** or expected variation for this variable?

3. What is the general **shape**, e.g., is it symmetric or skewed in some way?

For more background on these questions, see `https://www.itl.nist.gov/div898/hand book/ppc/section1/ppc131.htm`

This summary processing will become part of an automated acquire, clean, and summarize operation. The **User Experience (UX)** will be a command-line application. Our expected command line should look something like the following:

```
% python src/summarize.py -o summary/series_1/2023/03
data/clean/Series_1.ndj
```

The `-o` option specifies the path to an output sub-directory. The output filename added to this path will be derived from the source file name. The source file name often encodes information on the applicable date range for the extracted data.

> The Anscombe's Quartet data doesn't change and wouldn't really have an "applicable date" value.
>
> We've introduced the **idea** of periodic enterprise extractions. None of the projects actually specify a data source subject to periodic change.
>
> Some web services like `http://www.yelp.com` have health-code data for food-service businesses; this is subject to periodic change and serves as a good source of analytic data.

Now that we've seen the expectations, we can turn to an approach to the implementation.

Approach

We'll take some guidance from the C4 model (`https://c4model.com`) when looking at our approach:

- **Context**: For this project, a context diagram would show a user creating analytical reports. You may find it helpful to draw this diagram.

- **Containers**: There only seems to be one container: the user's personal computer.

- **Components**: We'll address the components below.

- **Code**: We'll touch on this to provide some suggested directions.

The heart of this application is a module to summarize data in a way that lets us test whether it fits the expectations of a model. The statistical model is a simplified reflection of the underlying real-world processes that created the source data. The model's simplifications include assumptions about events, measurements, internal state changes, and other details of the processing being observed.

For very simple cases — like Anscombe's Quartet data — there are only two variables, which leaves a single relationship in the model. Each of the four sample collections in the quartet has a distinct relationship. Many of the summary statistics, however, are the same, making the relationship often surprising.

For other datasets, with more variables and more relationships, there are numerous choices available to the analyst. The *NIST Engineering Statistics Handbook* has an approach to modeling. See `https://www.itl.nist.gov/div898/handbook/index.htm` for the design of a model and analysis of the results of the model.

As part of the preliminary work, we will distinguish between two very broad categories of statistical summaries:

- **Univariate statistics**: These are variables viewed in isolation.

- **Multivariate statistics**: These are variables in pairs (or higher-order groupings) with an emphasis on the relationship between the variable's values.

For univariate statistics, we need to understand the distribution of the data. This means measuring the location (the center or expected values), the spread (or scale), and the shape of the distribution. Each of these measurement areas has several well-known statistical functions that can be part of the summary application.

We'll look at the multivariate statistics in the next chapter. We'll start the univariate processing by looking at the application in a general way, and then focus on the statistical measures, the inputs, and finally, the outputs.

Designing a summary app

This application has a command-line interface to create a summary from the cleaned data. The input file(s) are the samples to be summarized. The summary must be in a form that's easy to process by subsequent software. This can be a JSON- or a TOML-formatted file with the summary data.

The summaries will be "measures of location," sometimes called a "central tendency." See `https://www.itl.nist.gov/div898/handbook/eda/section3/eda351.htm`.

The output must include enough context to understand the data source, and the variable being measured. The output also includes the measured values to a sensible number of decimal places. It's important to avoid introducing additional digits into floating-point values when those digits are little more than noise.

A secondary feature of this application is to create an easy-to-read presentation of the summary. This can be done by using tools like **Docutils** to transform a reStructuredText report into HTML or a PDF. A tool like **Pandoc** could also be used to convert a source report into something that isn't simply text. The technique explored in *Chapter 14, Project 4.2: Creating Reports* is to use Jupyter{Book} to create a document suitable for publication.

We'll start by looking at some of the measures of location that need to be computed.

Describing the distribution

As noted above, there are three aspects of the distribution of a variable. The data tends to scatter around a central tendency value; we'll call this the location. There will be an expected limit on the scattering; we'll call this the spread. There may be a shape that's symmetric or skewed in some way. The reasons for scattering may include measurement variability, as well as variability in the process being measured.

The NIST Handbook defines three commonly-used measures of location:

- **mean**: The sum of the variable's values divided by the count of values: $\bar{X} = \frac{\sum X_i}{N}$.

- **median**: The value of a value that is in the center of the distribution. Half the values are less than or equal to this value, and half the values are greater than or equal to this value. First, sort the values into ascending order. If there's an odd number, it's the value in the center. For an even number of values, split the difference between the two center-most values.

- **mode**: The most common value. For some of the Anscombe Quartet data series, this isn't informative because all of the values are unique.

These functions are first-class parts of the built-in `statistics` module, making them relatively easy to compute.

There are some alternatives that may be needed when the data is polluted by outliers. There are techniques like *Mid-Mean* and *Trimmed Mean* to discard data outside some range of percentiles.

The question of an "outlier" is a sensitive topic. An outlier may reflect a measurement problem. An outlier may also hint that the processing being measured is quite a bit more complicated than is revealed in a set of samples. Another, separate set of samples may reveal a different mean or a larger standard deviation. The presence of outliers may suggest more study is needed to understand the nature of these values.

There are three commonly-used measures for the scale or spread of the data:

- **Variance** and standard deviation. The variance is — essentially — the average of the squared distance of each sample from the mean: $s^2 = \frac{\sum X_i - \bar{X}}{(N-1)}$. The standard deviation is the square root of the variance.

- **Range** is the difference between the largest and smallest values.

- **Median absolute deviation** is the median of the distance of each sample from the mean: $\mathrm{MAD}_Y = \mathrm{median}(|Y_i - \tilde{Y}|)$. See *Chapter 7, Data Inspection Features*.

The variance and standard deviation functions are first-class parts of the built-in `statistics` module. The range can be computed using the built-in `min()` and `max()` functions. A median absolute deviation function can be built using functions in the `statistics` module.

There are also measures for skewness and kurtosis of a distribution. We'll leave these as extras to add to the application once the base statistical measures are in place.

Use cleaned data model

It's essential to use the cleaned, normalized data for this summary processing. There is some overlap between an inspection notebook and this more detailed analysis. An initial inspection may also look at some measures of location and range to determine if the data can be used or contains errors or problems. During the inspection activities, it's common to start creating an intuitive model of the data. This leads to formulating hypotheses about the data and considering experiments to confirm or reject those hypotheses.

This application formalizes hypothesis testing. Some functions from an initial data inspection notebook may be refactored into a form where those functions can be used on the cleaned data. The essential algorithm may be similar to the raw data version of the function. The data being used, however, will be the cleaned data.

This leads to a sidebar design decision. When we look back at the data inspection notebook, we'll see some overlaps.

Rethink the data inspection functions

Because Python programming can be generic — independent of any specific data type — it's tempting to try to unify the raw data processing and the cleaned data processing. The desire manifests as an attempt to write exactly one version of some algorithm, like the Median Absolute Deviation function that's usable for *both* raw and cleaned data.

This is not always an achievable goal. In some situations, it may not even be desirable.

A function to process raw data must often do some needed cleaning and filtering. These overheads are later refactored and implemented in the pipeline to create cleaned data. To be very specific, the if conditions used to exclude bad data can be helpful during the inspection. These conditions will become part of the clean-and-convert applications. Once this is done, they are no longer relevant for working with the cleaned data.

Because the extra data cleanups are required for inspecting raw data, but not required for analyzing cleaned data, it can be difficult to create a single process that covers both cases. The complications required to implement this don't seem to be worth the effort.

There are some additional considerations. One of these is the general design pattern followed by Python's `statistics` module. This module works with sequences of atomic values. Our applications will read (and write) complicated `Sample` objects that are not atomic Python integer or float values. This means our applications will extract sequences of atomic values from sequences of complicated `Sample` objects.

The raw data, on the other hand, may not have a very sophisticated class definition. This means the decomposition of complicated objects isn't part of the raw data processing.

For some very, very large datasets the decomposition of complicated multivariate objects to individual values may happen as the data is being read. Rather than ingest millions of objects, the application may extract a single attribute for processing.

This might lead to input processing that has the following pattern:

```python
from collections.abc import Iterator, Callable
import json
from pathlib import Path
from typing import TypeAlias


from analysis_model import Sample


Extractor: TypeAlias = Callable[[Sample], float]


def attr_iter(some_path: Path, extractor: Extractor) -> Iterator[float]:
    with some_path.open() as source:
        for line in source:
            document = Sample(**json.loads(line))
            yield extractor(document)


def x_values(some_path: Path) -> list[float]:
    return list(attr_iter(some_path, lambda sample: sample.x))
```

This example defines a generic function, `attr_iter()`, to read an ND JSON file to build instances of some class, `Sample`. (The details of the `Sample` class are omitted.)

The `x_values()` function uses the generic `attr_iter()` function with a concrete lambda object to extract a specific variable's value, and create a list object. This list object can then be used with various statistical functions.

While a number of individual `Sample` objects are created, they aren't retained. Only the values of the x attribute are saved, reducing the amount of memory used to create summary statistics from a large collection of complicated values.

Create new results model

The statistical summary contains three broad kinds of data:

- Metadata to specify what source data is used to create the summary.

- Metadata to specify what measures are being used.

- The computed values for location, shape, and spread.

In some enterprise applications, source data is described by a range of dates defining the earliest and latest samples. In some cases, more details are required to describe the complete context. For example, the software to acquire raw data may have been upgraded in the past. This means older data may be incomplete. This means the context for processing data may require some additional details on software versions or releases in addition to the range of dates and data sources.

Similarly, the measures being used may shift over time. The computation of skewness, for example, may switch from the **Fisher-Pearson** formula to the **adjusted Fisher-Pearson** formula. This suggests the version information for the summary program should also be recorded along with the results computed.

Each of these metadata values provides necessary context and background information on the data source, the method of collection, and any computations of derived data. This context may be helpful in uncovering the root cause of changes. In some cases, the context is a way to catalog underlying assumptions about a process or a measurement instrument; seeing this context may allow an analyst to challenge assumptions and locate the root cause of a problem.

The application must create a result document that looks something like the following example:

```
[identification]
    date = "2023-03-27T10:04:00"
[creator]
    title = "Some Summary App"
    version = 4.2
[source]
    title = "Anscombe's Quartet"
    path = "data/clean/Series_1.ndj"
[x.location]
    mean = 9.0
[x.spread]
    variance = 11.0
[y.location]
    mean = 7.5
[y.spread]
    variance = 4.125
```

This file can be parsed by the `toml` or `tomllib` module to create a nested collection of dictionaries. The secondary feature of the summary application is to read this file and write a report, perhaps using Markdown or ReStructuredText that provides the data in a readable format suitable for publication.

For Python 3.11 or newer, the `tomllib` module is built in. For older Python installations, the `toml` module needs to be installed.

Now that we've seen the overall approach, we can look at the specific deliverable files.

Deliverables

This project has the following deliverables:

- Documentation in the docs folder.

- Acceptance tests in the tests/features and tests/steps folders.

- Unit tests for model module classes in the tests folder.

- Mock objects for the csv_extract module tests will be part of the unit tests.

- Unit tests for the csv_extract module components that are in the tests folder.

- An application to summarize the cleaned data in a TOML file.

- An application secondary feature to transform the TOML file to an HTML page or PDF file with the summary.

In some cases, especially for particularly complicated applications, the summary statistics may be best implemented as a separate module. This module can then be expanded and modified without making significant changes to the overall application.

The idea is to distinguish between these aspects of this application:

- The CLI, which includes argument parsing and sensible handling of input and output paths.

- The statistical model, which evolves as our understanding of the problem domain and the data evolve.

- The data classes, which describe the structure of the samples, independent of any specific purpose.

For some applications, these aspects do not involve a large number of classes or functions. In a case where the definitions are small, a single Python module will do nicely. For other applications, particularly those where initial assumptions turned out to be invalid and significant changes were made, having separate modules can permit more flexibility, and more agility with respect to future changes.

We'll look at a few of these deliverables in a little more detail. We'll start with some suggestions for creating the acceptance tests.

Acceptance testing

The acceptance tests need to describe the overall application's behavior from the user's point of view. The scenarios will follow the UX concept of a command-line application to acquire data and write output files. Because the input data has been cleaned and converted, there are few failure modes for this application; extensive testing of potential problems isn't as important as it was in earlier data-cleaning projects.

For relatively simple datasets, the results of the statistical summaries are known in advance. This leads to features that might look like the following example:

```
Feature: Summarize an Anscombe Quartet Series.


Scenario: When requested, the application creates a TOML summary of a series.
  Given the "clean/series_1.ndj" file exists
  When we run command "python src/summarize.py \
        -o summary/series_1/2023/03 data/clean/Series_\1.ndj"
  Then the "summary/series_1/2023/03/summary.toml" file exists
  And the value of "summary['creator']['title']" is "Anscombe Summary App"
  And the value of "summary['source']['path']" is "data/clean/Series_1.ndj"
  And the value of "summary['x']['location']['mean']" is "9.0"
```

We could continue the scenario with a number of additional Then steps to validate each of the locations and the spread and shape the statistical summaries.

The step definitions will be similar to step definitions for a number of previous projects. Specifically, the When step will use the subprocess.run() function to execute the given application with the required command-line arguments.

The first of the Then steps will need to read — and parse — the TOML file. The resulting summary object can be placed in the context object. Subsequent Then steps can examine the structure to locate the individual values, and confirm the values match the acceptance test expectations.

 It is often helpful to extract a small subset of data to use for acceptance testing. Instead of processing millions of rows, a few dozen rows are adequate to confirm the application has read and summarized data. The data only needs to be representative of the larger set of samples under consideration.

Because the chosen subset is part of the testing suite; it rarely changes. This makes the results predictable.

As the data collection process evolves, it's common to have changes to the data sources. This will lead to changes in the data cleaning. This may, in turn, lead to changes in the summary application as new codes or new outliers must be handled properly. The evolution of the data sources implies that the test data suite will also need to evolve to expose any of the special, edge, or corner cases.

Ideally, the test data suite is a mixture of ordinary — no surprises — data, mixed with representative examples of each of the special, atypical cases. As this test data suite evolves, the acceptance test scenario will also evolve.

The TOML file is relatively easy to parse and verify. The secondary feature of this application — expanding on the TOML output to add extensive Markdown — also works with text files. This makes it relatively easy to confirm with test scenarios that read and write text files.

The final publication, whether done by Pandoc or a combination of Pandoc and a LaTeX toolchain, isn't the best subject for automated testing. A good copy editor or trusted associate needs to make sure the final document meets the stakeholder's expectations.

Unit testing

It's important to have unit testing for the various components that are unique to this application. The clean data class definition, for example, is created by another application, with its own test suite. The unit tests for this application don't need to repeat those tests. Similarly, the `statistics` module has extensive unit tests; this application's unit tests do not need to replicate any of that testing.

This further suggests that the `statistics` module should be replaced with `Mock` objects. Those mock objects can — generally — return `sentinel` objects that will appear in the resulting TOML-format summary document.

This suggests test cases structured like the following example:

```python
from pytest import fixture
from unittest.mock import Mock, call, sentinel
import summary_app


@fixture
def mocked_mean(monkeypatch):
    mean = Mock(
        return_value=sentinel.MEAN
    )
    monkeypatch.setattr(summary_app, 'mean', mean)
    return mean


@fixture
def mocked_variance(monkeypatch):
    variance = Mock(
        return_value=sentinel.VARIANCE
    )
    monkeypatch.setattr(summary_app, 'variance', variance)
    return variance


def test_var_summary(mocked_mean, mocked_variance):
    sample_data = sentinel.SAMPLE
```

```
result = summary_app.variable_summary(sample_data)
assert result == {
    "location": {"mean": sentinel.MEAN},
    "spread": {"variance": sentinel.VARIANCE},
}
assert mocked_mean.mock_calls == [call(sample_data)]
assert mocked_variance.mock_calls == [call(sample_data)]
```

The two test fixtures provide mock results, using `sentinel` objects. Using `sentinel` objects allows easy comparison to be sure the results of the mocked functions were not manipulated unexpectedly by the application.

The test case, `test_var_summary()`, provides a mocked source of data in the form of another `sentinel` object. The results have the expected structure and the expected `sentinel` objects.

The final part of the test confirms the sample data — untouched — was provided to the mocked statistical functions. This confirms the application doesn't filter or transform the data in any way. The results are the expected `sentinel` objects; this confirms the module didn't adulterate the results of the `statistics` module. And the final check confirms that the mocked functions were called exactly once with the expected parameters.

This kind of unit test, with numerous mocks, is essential for focusing the testing on the new application code, and avoiding tests of other modules or packages.

Application secondary feature

A secondary feature of this application transforms the TOML summary into a more readable HTML or PDF file. This feature is a variation of the kinds of reporting done with Jupyter Lab (and associated tools like **Jupyter {Book}**).

There's an important distinction between these two classes of reports:

- The Jupyter Lab reports involve discovery. The report content is always new.

- The summary application's reports involve confirmation of expectations. The report content should not be new or surprising.

In some cases, the report will be used to confirm (or deny) an expected trend is continuing. The application applies the trend model to the data. If the results don't match expectations, this suggests follow-up action is required. Ideally, it means the model is incorrect, and the trend is changing. The less-than-ideal case is the observation of an unexpected change in the applications providing the source data.

This application decomposes report writing into three distinct steps:

1. **Content**: This is the TOML file with the essential statistical measures.

2. **Structure**: The secondary feature creates an intermediate markup file in Markdown or the RST format. This has an informative structure around the essential content.

3. **Presentation**: The final publication document is created from the structured markup plus any templates or style sheets that are required.

The final presentation is kept separate from the document's content and structure.

 An HTML document's final presentation is created by a browser. Using a tool like **Pandoc** to create HTML from Markdown is — properly — replacing one markup language with another markup language.

Creating a PDF file is a bit more complicated. We'll leave this in the extras section at the end of this chapter.

The first step toward creating a nicely formatted document is to create the initial Markdown or ReStructuredText document from the summary. In many cases, this is easiest done with the **Jinja** package. See `https://jinja.palletsprojects.com/en/3.1.x/`

One common approach is the following sequence of steps:

1. Write a version of the report using Markdown (or RST).

2. Locate a template and style sheets that produce the desired HTML page when converted by the **Pandoc** or **Docutils** applications.

3. Refactor the source file to replace the content with **Jinja** placeholders.

This becomes the template report.

4. Write an application to parse the TOML, then apply the TOML details to the template file.

When using **Jinja** to enable filling in the template, it must be added to the `requirements.txt` file. If **ReStructuredText (RST)** is used, then the **docutils** project is also useful and should be added to the `requirements.txt` file.

If Markdown is used to create the report, then **Pandoc** is one way to handle the conversion from Markdown to HTML. Because **Pandoc** also converts RST to HTML, the **docutils** project is not required.

Because the parsed TOML is a dictionary, fields can be extracted by the Jinja template. We might have a Markdown template file with a structure like the following:

```
# Summary of {{ summary['source']['name'] }}


Created {{ summary['identification']['date'] }}


Some interesting notes about the project...


## X-Variable


Some interesting notes about this variable...


Mean = {{ summary['x']['location']['mean'] }}


etc.
```

The `{{ some-expression }}` constructs are placeholders. This is where Jinja will evaluate the Python expression and replace the placeholders with the resulting value. Because of Jinja's clever implementation, a name like `summary['x']['location']['mean']` can be written as `summary.x.location.mean`, also.

The lines with # and ## are the way Markdown specifies the section headings. For more information on Markdown, see `https://daringfireball.net/projects/markdown/`. Note that there are a large number of Markdown extensions, and it's important to be sure the rendering engine (like Pandoc) supports the extensions you'd like to use.

The Jinja template language has numerous options for conditional and repeating document sections. This includes `{% for name in sequence %}` and `{% if condition %}` constructs to create extremely sophisticated templates. With these constructs, a single template can be used for a number of closely related situations with optional sections to cover special situations.

The application program to inject values from the `summary` object into the template shouldn't be much more complicated than the examples shown on the Jinja basics page. See `https://jinja.palletsprojects.com/en/3.1.x/api/#basics` for some applications that load a template and inject values.

This program's output is a file with a name like `summary_report.md`. This file would be ready for conversion to any of a large number of other formats.

The process of converting a Markdown file to HTML is handled by the **Pandoc** application. See `https://pandoc.org/demos.html`. The command might be as complicated as the following:

```
pandoc -s --toc -c pandoc.css summary_report.md -o summary_report.html
```

The `pandoc.css` file can provide the CSS styles to create a body that's narrow enough to be printed on an ordinary US letter or A4 paper.

The application that creates the `summary_report.md` file can use `subprocess.run()` to execute the **Pandoc** application and create the desired HTML file. This provides a command-line UX that results in a readable document, ready to be distributed.

Summary

In this chapter we have created a foundation for building and using a statistical model of source data. We've looked at the following topics:

- Designing and building a more complex pipeline of processes for gathering and analyzing data.

- Some of the core concepts behind creating a statistical model of some data.

- Use of the built-in `statistics` library.

- Publishing the results of the statistical measures.

This application tends to be relatively small. The actual computations of the various statistical values leverage the built-in `statistics` library and tend to be very small. It often seems like there's far more programming involved in parsing the CLI argument values, and creating the required output file, than doing the "real work" of this application.

This is a consequence of the way we've been separating the various concerns in data acquisition, cleaning, and analysis. We've partitioned the work into several, isolated stages along a pipeline:

1. Acquiring raw data, generally in text form. This can involve database access or RESTful API access, or complicated file parsing problems.

2. Cleaning and converting the raw data to a more useful, native Python form. This can involve complications of parsing text and rejecting outlier values.

3. Summarizing and analyzing the cleaned data. This can focus on the data model and reporting conclusions about the data.

The idea here is the final application can grow and adapt as our understanding of the data matures. In the next chapter, we'll add features to the summary program to create deeper insights into the available data.

Extras

Here are some ideas for you to add to this project.

Measures of shape

The measurements of shape often involve two computations for skewness and kurtosis. These functions are not part of Python's built-in `statistics` library.

It's important to note that there are a very large number of distinct, well-understood distributions of data. The normal distribution is one of many different ways data can be distributed.

See `https://www.itl.nist.gov/div898/handbook/eda/section3/eda366.htm`.

One measure of skewness is the following:

$$g_1 = \frac{\frac{\sum(Y_i - \bar{Y})^3}{N}}{s^3}$$

Where $\bar{Y}$ is the mean, and s is the standard deviation.

A symmetric distribution will have a skewness, g_1, near zero. Larger numbers indicate a "long tail" opposite a large concentration of data around the mean.

One measure of kurtosis is the following:

$$\text{kurtosis} = \frac{\frac{\sum(Y_i - \bar{Y})^4}{N}}{s^4}$$

The kurtosis for the standard normal distribution is 3. A value larger than 3 suggests more data is in the tails; it's "flatter" or "wider" than the standard normal distribution. A value less than three is "taller" or "narrower" than the standard.

These metrics can be added to the application to compute some additional univariate descriptive statistics.

Creating PDF reports

In the *Application secondary feature* section we looked at creating a Markdown or RST document with the essential content, some additional information, and an organizational structure. The intent was to use a tool like **Pandoc** to convert the Markdown to HTML. The HTML can be rendered by a browser to present an easy-to-read summary report.

Publishing this document as a PDF requires a tool that can create the necessary output file. There are two common choices:

- Use the **ReportLab** tool: `https://www.reportlab.com/dev/docs/`. This is a commercial product with some open-source components.

- Use the **Pandoc** tool coupled with a LaTeX processing tool.

See *Preparing a report* of *Chapter 14, Project 4.2: Creating Reports* for some additional thoughts on using LaTeX to create PDF files. While this involves a large number of separate components, it has the advantage of having the most capabilities.

It's often best to learn the LaTeX tools separately. The TeXLive project maintains a number of tools useful for rendering LaTeX. For macOS users, the MacTex project offers the required binaries. An online tool like Overleaf is also useful for handling LaTeX. Sort out any problems by creating small `hello_world.tex` example documents to see how the LaTeX tools work.

Once the basics of the LaTeX tools are working, it makes sense to add the **Pandoc** tool to the environment.

Neither of these tools are Python-based and don't use **conda** or **pip** installers.

As noted in *Chapter 14, Project 4.2: Creating Reports*, there are a lot of components to this tool chain. This is a large number of separate installs that need to be managed. The results, however, can be very nice when a final PDF is created from a few CLI interactions.

Serving the HTML report from the data API

In *Chapter 12, Project 3.8: Integrated Data Acquisition Web Service* we created a RESTful API service to provide cleaned data.

This service can be expanded to provide several other things. The most notable addition is the HTML summary report.

The process of creating a summary report will look like this:

1. A user makes a request for a summary report for a given time period.

2. The RESTful API creates a "task" to be performed in the background and responds with the status showing the task has been created.

3. The user checks back periodically to see if the processing has finished. Some clever JavaScript programming can display an animation while an application program checks to see if the work is completed.

4. Once the processing is complete, the user can download the final report.

This means two new resources paths will need to be added to the OpenAPI specification. These two new resources are:

- Requests to create a new summary. A POST request creates the task to build a summary and a GET request shows the status. A `2023.03/summarize` path will parallel the `2023.02/creation` path used to create the series.

- Requests for a summary report. A GET request will download a given statistical summary report. Perhaps a `2023.03/report` path would be appropriate.

As we add features to the RESTful API, we need to consider the resource names more and more carefully. The first wave of ideas sometimes fails to reflect the growing understanding of the user's needs.

In retrospect, the 2023.02/create path, defined in *Chapter 12, Project 3.8: Integrated Data Acquisition Web Service,* may not have been the best name.

 There's an interesting tension between requests to create a resource and the resulting resource. The request to create a series is clearly distinct from the resulting series. Yet, they can both be meaningfully thought of as instances of "series." The creation request is a kind of *future*: an expectation that will be fulfilled later.

An alternative naming scheme is to use 2023.02/creation for series, and use 2023.03/create/series and 2023.03/create/summary as distinct paths to manage the long-running background that does the work.

The task being performed in the background will execute a number of steps:

1. Determine if the request requires new data or existing data. If new data is needed, it is acquired, and cleaned. This is the existing process to acquire the series of data points.

2. Determine if the requested summary does not already exist. (For new data, of course, it will not exist.) If a summary is needed, it is created.

Once the processing is complete, the raw data, cleaned data, and summary can all be available as resources on the API server. The user can request to download any of these resources.

It's essential to be sure each of the components for the task work in isolation before attempting to integrate them as part of a web service. It's far easier to diagnose and debug problems with summary reporting outside the complicated world of web services.

16
Project 5.2: Simple Multivariate Statistics

Are variables related? If so what's the relationship? An analyst tries to answer these two questions. A negative answer — the null hypothesis — doesn't require too many supporting details. A positive answer, on the other hand, suggests that a model can be defined to describe the relationship. In this chapter, we'll look at simple correlation and linear regression as two elements of modeling a relationship between variables.

In this chapter, we'll expand on some skills of data analysis:

- Use of the built-in `statistics` library to compute correlation measures and linear regression coefficients.

- Use of the **matplotlib** library to create images. This means creating plot images outside a Jupyter Lab environment.

- Expanding on the base modeling application to add features.

This chapter's project will expand on earlier projects. Look back at *Chapter 13, Project 4.1: Visual Analysis Techniques* for some of the graphical techniques used in a Jupyter Lab context. These need to be more fully automated. The project will add multivariate statistics and graphs to illustrate relationships among variables.

Description

In *Chapter 15, Project 5.1: Modeling Base Application* we created an application to create a summary document with some core statistics. In that application, we looked at univariate statistics to characterize the data distributions. These statistics included measurements of the location, spread, and shape of a distribution. Functions like mean, median, mode, variance, and standard deviation were emphasized as ways to understand location and spread. The characterization of shape via skewness and kurtosis was left as an extra exercise for you.

The base application from the previous chapter needs to be expanded to include the multivariate statistics and diagrams that are essential for clarifying the relationships among variables. There are a vast number of possible functions to describe the relationships among two variables. See `https://www.itl.nist.gov/div898/handbook/pmd/section8/pmd8.htm` for some insight into the number of choices available.

We'll limit ourselves to linear functions. In the simplest cases, there are two steps to creating a linear model: identifying a correlation and creating the coefficients for a line that fits the data. We'll look at each of these steps in the next two sections.

Correlation coefficient

The coefficient of correlation measures how well the values of two variables correlate with each other. A value of 1 indicates perfect correlation; a value of zero indicates no discernable correlation. A value of -1 indicates an "anti-correlation": when one variable is at its maximum value, the other variable is at its minimum.

See *Figure 16.1* to see how the correlation coefficient describes the distribution of the two variables.

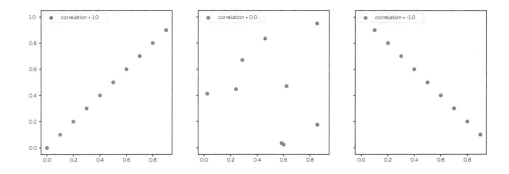

Figure 16.1: Correlation Coefficients

The computation of the coefficient compares individual values of variables, X_i and Y_i, to the mean values for those variables, $\bar{X}$ and $\bar{Y}$. Here's a formula:

$$r = \frac{\sum(X_i - \bar{X})(Y_i - \bar{Y})}{\sqrt{\sum(X_i - \bar{X})^2}\sqrt{\sum(Y_i - \bar{Y})^2}}$$

The computations of the mean values, $\bar{X}$ and $\bar{Y}$, can be factored into this, creating a somewhat more complicated version that's often used to create the coefficient in a single pass through the data.

This function is available as `statistics.correlation()` in the standard library.

If two variables correlate with each other, then a linear function will map one variable to a value near the other variable. If the correlation is 1.0 or -1.0, the mapping will be exact. For other correlation values, the mapping will be close, but not exact. In the next section, we'll show how to transform the correlation coefficient into the parameters for a line.

Linear regression

One equation for a line is $y = mx + b$. The values of m and b are parameters that describe the specific linear relationship between the x and y variables.

When fitting a line to data, we're estimating the parameters for a line. The goal is to minimize the error between the line and the actual data. The "least squares" technique is often used.

The two coefficients, b and m, can be computed as follows:

$$b = \frac{\sum X_i^2 \sum Y_i^2 - \sum X_i \sum X_i Y_i}{n \sum X_i^2 - (\sum X_i)^2}$$

$$m = \frac{n \sum X_i Y_i - \sum X_i \sum Y_i}{n \sum X_i^2 - (\sum X_i)^2}$$

This is available as `statistics.linear_regression()` in the standard library. This saves us from having to write these two functions.

The various sums and sums of squares are not terribly difficult values to compute. The built-in `sum()` function is the basis for most of this. We can use `sum(map(lambda x: x^2, x_values))` to compute $\sum X_i^2$.

To clarify these multivariate relationships, diagrams can be very helpful. In the next sections, we'll look at the most important type of diagram that needs to be part of the overall application.

Diagrams

One essential diagram for showing multivariate data is the X-Y "scatter" plot. In *Chapter 13, Project 4.1: Visual Analysis Techniques* we looked at ways to create these. In that chapter, we relied on Jupyter Lab to present the diagram as part of the overall web page. For this application, we'll need to embed the diagram into a document.

This generally means there will be a markup document that includes a reference to a diagram file. The format of the diagram file can be SVG, PNG, or even JPEG. For technical graphics, the SVG files are often the smallest and scale extremely well.

Each markup language, including Markdown, RST, HTML, and LaTeX, have unique ways to identify the place where an image needs to be inserted. In the case of Markdown, it's often necessary to use HTML syntax to properly include frames and captions.

Now that we've seen what the application needs to do, we can look at an approach to create the software.

Approach

As with the previous project, this application works in these two distinct parts:

1. Compute the statistics and create the diagram files.

2. Create a report file in a simplified markup language from a template with the details interpolated. A tool like **Jinja** is very helpful for this.

Once the report file in a markup language — like Markdown or RST — is available, then a tool like **Pandoc** can be used to create an HTML page or a PDF document from the markup file. Using a tool like **Pandoc** permits quite a bit of flexibility in choosing the final format. It also allows the insertion of style sheets and page templates in a tidy, uniform way.

> The LaTeX language as markup provides the most comprehensive capabilities. It is challenging to work with, however. Languages like Markdown and RST are designed to offer fewer, easier-to-use capabilities.
>
> This book is written with LaTeX.

We'll look at three aspects of this application: the statistical computations, creating the diagrams, and finally, creating the final markup file to include the diagrams. We'll start with a quick review of the statistical computations.

Statistical computations

The statistical summary output file, in TOML notation, has a section for each variable and the univariate statistics about those variables.

This section of the file looked like the following snippet of TOML:

```
[x.location]
    mean = 9.0
[x.spread]
    variance = 11.0
[y.location]
    mean = 7.5
[y.spread]
    variance = 4.125
```

When parsed, the TOML syntax of x.location and x.spread creates a dictionary that looks like the following fragment of a Python object:

```
{
    some metadata here...

    'x': {
        'location': {
            'mean': 9.0
        },
        'spread': {
            'variance': 11.0
        }
    },
    'y': {
        etc.
    }
}
```

This structure can be expanded to include additional locations and spread statistical measures. It can also be expanded to include multivariate statistics.

The `statistics` module has `correlation()` and `covariance()` functions, making it easy to include these measures.

For datasets with few variables, it's common to consider a matrix that includes all the combinations of covariance between variables. This leads to two alternative representations of these additional statistics:

- A separate section for a covariance matrix. A section label of [covariance] can be followed by nested dictionaries that include all combinations of variables. Since the covariance matrix is symmetric, all n^2 combinations aren't needed; only $n \times (n-1)$ values are unique.

- Multivariate sub-sections within each variable's section. This means we'd have `x.location`, `x.spread`, `x.covariance.y`, and `x.correlation.y` sub-sections for the x variable.

For a dataset with fewer variables, it seems sensible to bundle covariance and correlation into the details for a given variable. In the case of Anscombe's Quartet, with only two variables, the covariance and correlation seem like they belong with the other statistics.

For a dataset with a larger number of variables, the covariance among all the variables can become bewildering. In these cases, a technique like finding principal components might be needed to reduce the number of variables to a more manageable population. In this case, separate sections with covariance and auto-correlation might be more useful.

The resulting model is often the result of some careful thought, based on the covariance matrix. For this reason, a separate [model] section should be provided with some details about the model's structure and the coefficients. In the case of a linear model, there are two coefficients, sometimes called β_0 and β_1. We've called them b and m.

For Python 3.11, the included `tomllib` module doesn't create TOML-format files. It's, therefore, necessary to properly format a text file that can be parsed by the `tomllib` module. It's helpful to use a Jinja template for this.

Analysis diagrams

Diagrams must first be created. Once created, they can then be included in a document. The process of creating a diagram is nearly identical to the approach used in Jupyter Lab. A few extra steps need to be taken to export the diagram to a file that can be imported into a document.

When working in Jupyter Lab, some cells to load the data are required to create two variables, x and y, with the values to be plotted. After these cells, a cell like the following example will create and display a scatter plot:

```python
import matplotlib.pyplot as plt


fig, ax = plt.subplots()


# Labels and Title
ax.set_xlabel('X')
ax.set_ylabel('Y')
ax.set_title('Series I')


# Draw Scatter
_ = ax.scatter(x, y)
```

This presumes previous cells have loaded clean data and extracted two list objects, x and y, with the values to be plotted.

The above code sample doesn't save the resulting PNG or SVG file, however. To save the figure, we need to perform two more steps. Here are the lines of code required to create a file from the plot:

```python
    plt.savefig('scatter_x_y.png')
    plt.close(fig)
```

It helps to transform this cell's code into a function. This function has proper type annotations so that a tool like **mypy** can confirm the types are used properly. It can also have unit test cases to be sure it really works.

The `savefig()` function will write a new file in PNG format with the image. If the file path suffix is `'.svg'` then an SVG format file will be created.

The size of the figure is defined by the `figure()` function. There are often design and page layout considerations that suggest an appropriate size for a figure. This decision can be deferred, and the size can be provided by the markup used to create a final PDF file or HTML page. It's often best, however, to create the figure in the required size and resolution to avoid any unexpected alterations as part of the final publication.

Once the diagram has been created, the Markdown needs to refer to the diagram's PNG or SVG file so it can be included in a document. We'll look at some examples of this in the next section.

Including diagrams in the final document

Diagrams are included in the final document by using markup commands to show where the diagram should be placed, and providing other information about captions and sizing.

The Markdown language has a tidy format for the simplest case of including an image in a document:

```
![Alt text to include!](path/to/file.png "Figure caption")
```

Depending on the style sheet, this may be perfectly acceptable. In some cases, the image is the wrong size for its role in the document. Markdown permits using HTML directly instead of the `![image!](path)` construct. Including a diagram often looks like this:

```
<figure>
    <img src="path/to/file.png"
        alt="Alt text to include"
        height="8cm">
```

```
    <figcaption>Figure caption</figcaption>
</figure>
```

Using HTML permits more control over image size and placement via references to CSS.

When using RST, the syntax offers more options without switching to HTML. Including a diagram would be like this:

```
..  figure:: path/to/file.png
    :height: 8cm
    :alt: Alt text to include


    Figure caption
```

Using this kind of markup technique creates considerable freedom. The report's author can include content from a variety of sources. This can include boilerplate text that doesn't change, the results of computations, some text based on the computations, and important diagrams.

The formatting of the markup has little impact on the final document. The way a browser renders HTML depends on the markup and the style sheets, not the formatting of the source file. Similarly, when creating a PDF document, this is often done by LaTeX tools, which create the final document based on LaTeX settings in the document's preamble.

Now that we have an approach, we can look at the deliverable files that must be built.

Deliverables

This project has the following deliverables:

- Documentation in the docs folder.

- Acceptance tests in the tests/features and tests/steps folders.

- Unit tests for model module classes in the tests folder.

- Mock objects for the csv_extract module tests that will be part of the unit tests.

- Unit tests for the `csv_extract` module components that are in the `tests` folder.

- An application to extend the summary written to a TOML file, including figures with diagrams.

- An application secondary feature to transform the TOML file to an HTML page or PDF file with the summary.

We'll look at a few of these deliverables in a little more detail. We'll start with some suggestions for creating the acceptance tests.

Acceptance tests

As we noted in the previous chapter's section on acceptance testing, *Acceptance testing*, the output TOML document can be parsed and examined by the Then steps of a scenario. Because we're looking at Anscombe's Quartet data in the examples in this book, a subset of data for testing doesn't really make much sense. For any other dataset, a subset should be extracted and used for acceptance testing.

 It is often helpful to extract a small subset that's used for acceptance testing. Instead of processing millions of rows, a few dozen rows are adequate to confirm the application read and summarized data. The data should be representative of the entire set of samples under consideration.

This subset is part of the testing suite; as such, it rarely changes. This makes the results predictable.

The secondary feature of this application — expanding on the TOML output to add extensive Markdown — also works with text files. This makes it relatively easy to create scenarios to confirm the correct behavior by reading and writing text files. In many cases, the Then steps will look for a few key features of the resulting document. They may check for specific section titles or a few important keywords included in boilerplate text. Of course, the test scenario can check for substitution values that are computed and are part of the TOML summary.

The automated testing can't easily confirm that the document makes sense to prospective readers. It can't be sure the colors chosen for the figures make the relationships clear. For this kind of usability test, a good copy editor or trusted associate is essential.

Unit tests

A unit test for a function to create a figure can't do very much. It's limited to confirming that a PNG or SVG file was created. It's difficult for an automated test to "look" at the image to be sure it has a title, labels for the axes, and sensible colors.

It is important not to overlook the unit test cases that confirm output files are created. A figure that looks great in a Jupyter notebook will not get written to a file unless the CLI application saves the figure to a file.

For some applications, it makes sense to mock the plt package functions to be sure the application calls the right functions with the expected argument values. Note that a mocked version of plt.subplots() may need to return a tuple with several Mock objects.

We'll need to define a complex collection of mock objects to form the fixture for testing. The fixture creation can look like the following example:

```
@fixture
def mocked_plt_module(monkeypatch):
    fig_mock = Mock()
    ax_mock = Mock(
        set_xlabel=Mock(),
        set_ylabel=Mock(),
        set_tiutle=Mock(),
        scatter=Mock(),
    )
    plt_mock = Mock(
        subplots=Mock(
            return_value=(fig_mock, ax_mock)
```

```
        ),
        savefig=Mock(),
        close=Mock()
    )
    monkeypatch.setattr(summary_app, 'plt', plt_mock)
    return plt_mock, fig_mock, ax_mock
```

This fixture creates three mock objects. The plt_mock is a mock of the overall plt module; it defines three mock functions that will be used by the application. The fig_mock is a mock of the figure object returned by the subplots() function. The ax_mock is a mock of the axes object, which is also returned by the subplots() function. This mocked axes object is used to provide axis labels, and the title, and perform the scatter plot request.

This three-tuple of mock objects is then used by a test as follows:

```
def test_scatter(mocked_plt_module):
    plt_mock, fig_mock, ax_mock = mocked_plt_module
    summary_app.scatter_figure([sentinel.X], [sentinel.Y])


    assert plt_mock.subplots.mock_calls == [call()]
    assert plt_mock.savefig.mock_calls == [call('scatter_x_y.png')]
    assert plt_mock.close.mock_calls == [call(fig_mock)]
```

This test function evaluates the application's scatter_figure() function. The test function then confirms that the various functions from the plt module are called with the expected argument values.

The test can continue by looking at the calls to the ax_mock object to see if the labels and title requests were made as expected. This level of detail — looking at the calls to the axes object — may be a bit too fine-grained. These tests become very brittle as we explore changing titles or colors to help make a point more clearly.

The overall use of mock objects, however, helps make sure the application will create the needed file with an image.

Summary

In this chapter, we've extended the automated analysis and reporting to include more use of the built-in `statistics` library to compute correlation and linear regression coefficients. We've also made use of the **matplotlib** library to create images that reveal relationships among variables.

The objective of automated reporting is designed to reduce the number of manual steps and avoid places where omissions or errors can lead to unreliable data analysis. Few things are more embarrassing than a presentation that reuses a diagram from the previous period's data. It's far too easy to fail to rebuild one important notebook in a series of analysis products.

The level of automation needs to be treated with a great deal of respect. Once a reporting application is built and deployed, it must be actively monitored to be sure it's working and producing useful, informative results. The analysis job shifts from developing an understanding to monitoring and maintaining the tools that confirm — or reject — that understanding.

In the next chapter, we'll review the journey from raw data to a polished suite of applications that acquires, cleans, and summarizes the data.

Extras

Here are some ideas for you to add to this project.

Use pandas to compute basic statistics

The **pandas** package offers a robust set of tools for doing data analysis. The core concept is to create a `DataFrame` that contains the relevant samples. The `pandas` package needs to be installed and added to the `requirements.txt` file.

There are methods for transforming a sequence of `SeriesSample` objects into a `DataFrame`. The best approach is often to convert each of the **pydantic** objects into a dictionary, and build the dataframe from the list of dictionaries.

The idea is something like the following:

```
import pandas as pd
```

```
df = pd.DataFrame([dict(s) for s in series_data])
```

In this example, the value of series_data is a sequence of SeriesSample instances.

Each column in the resulting dataframe will be one of the variables of the sample. Given this object, methods of the DataFrame object produce useful statistics.

The corr() function, for example, computes the correlation values among all of the columns in the dataframe.

The cov() function computes the pairwise covariance among the columns in the dataframe.

Pandas doesn't compute the linear regression parameters, but it can create a wide variety of descriptive statistics.

See https://pandas.pydata.org for more information on Pandas.

In addition to a variety of statistics computations, this package is designed for interactive use. It works particularly well with Juypyter Lab. The interested reader may want to revisit *Chapter 13, Project 4.1: Visual Analysis Techniques* using Pandas instead of native Python.

Use the dask version of pandas

The **pandas** package offers a robust set of tools for doing data analysis. When the volume of data is vast, it helps to process parts of the dataset concurrently. The **Dask** project has an implementation of the **pandas** package that maximizes opportunities for concurrent processing.

The dask package needs to be installed and added to the requirements.txt file. This will include a pandas package that can be used to improve overall application performance.

Use numpy for statistics

The **numpy** package offers a collection of tools for doing high-performance processing on large arrays of data. These basic tools are enhanced with libraries for statistics and linear algebra among many, many other features. This package needs to be installed and added to the `requirements.txt` file.

The **numpy** package works with its own internal array type. This means the `SeriesSample` objects aren't used directly. Instead, a `numpy.array` object can be created for each of the variables in the source series.

The conversion might look like the following:

```
import numpy as np


x = np.array(s.x for s in series_data)
y = np.array(s.y for s in series_data)
```

In this example, the value of `series_data` is a sequence of `SeriesSample` instances.

It's also sensible to create a single multi-dimensional array. In this case, axis 0 (i.e. rows) will be the individual samples, and axis 1 (i.e. columns) will be the values for each variable of the sample.

An array has methods like `mean()` to return the mean of the values. When using a multi-dimensional array, it's essential to provide the `axis=0` parameter to ensure that the results come from processing the collection of rows:

```
import numpy as np


a = np.array([[s.x, s.y] for s in series_data])
print(f"means = {a.mean(axis=0)}")
```

See `https://numpy.org/doc/stable/reference/routines.statistics.html#`

Using the least squares technique to compute the coefficients for a line can be confusing. The least squares solver in **numpy.linalg** is a very general algorithm, which can be applied to creating a linear model. The `numpy.linalg.lstsq()` function expects a small matrix that contains the "x" values. The result will be a vector with the same length as each of the "x" matrices. The "y" values will also be a vector.

The processing winds up looking something like the following:

```python
import numpy as np


A = np.array([[s.x, 1] for s in series_data])
y = np.array([s.y for s in series_data])
m, b = np.linalg.lstsq(A, y, rcond=None)[0]
print(f"y = {m:.1f}x + {b:.1f}")
```

The value of A is a small matrix based on the x values. The value of y is a simple array of the y values. The least-squares algorithm returns a four-tuple with the coefficients, residuals, the rank of the source matrix, and any singular values. In the above example, we only wanted the vector of the coefficients, so we used [0] to extract the coefficient values from the four-tuple with the results.

This is further decomposed to extract the two coefficients for the line that best fits this set of points. See:

`https://numpy.org/doc/stable/reference/generated/numpy.linalg.lstsq.html`.

This approach has a distinct advantage when working with very large sets of data. The **numpy** libraries are very fast and designed to scale to extremely large data volumes.

Use scikit-learn for modeling

The **scikit-learn** library has a vast number of tools focused on modeling and machine learning. This library is built on the foundation of **numpy**, so both packages need to be installed.

The data needs to be converted into **numpy** arrays. Because the modeling approach is very generalized, the assumption is there may be multiple independent variables that predict the value of a dependent variable.

The conversion might look like the following:

```
import numpy as np


x = np.array([[s.x] for s in series_data])
y = np.array([s.y for s in series_data])
```

In this example, the value of `series_data` is a sequence of `SeriesSample` instances. The x array uses a very short vector for each sample; in this case, there's only a single value. It needs to be a vector to fit the generalized least-squares regression that **scikit-learn** is capable of solving.

The scikit-learn library is designed to create models in a very generalized way. The model isn't always a simple line with a coefficient and an intercept that define the relationship. Because of this very general approach to modeling, we'll create an instance of the `linear_model.LinearRegression` class. This object has methods to create coefficients that fit a given set of data points. We can then examine the coefficients, or use them to interpolate new values.

The code might look like the following:

```
from sklearn import linear_model


reg = linear_model.LinearRegression()
reg.fit(x, y)
print(f"y = {reg.coef_[0]:.1f}x + {reg.intercept_:.1f}")
```

The linear model's `coef_` attribute is a vector of coefficients, the same length as each row of the x independent variable values. Even when the row length is 1, the result is a vector with a length of 1.

Because this works with **numpy** it can work with very large sets of data. Further, the scikit-learn approach to creating models to fit data generalizes to a number of machine-learning approaches. This is often the next step in creating richer and more useful models.

Compute the correlation and regression using functional programming

The computations for correlation and the coefficients for a line can be summarized as follows. First, we'll define a function $M(a; f())$ that computes the mean of a transformed sequence of values. The $f()$ function transforms each value, a_i. An identity function, $\phi(a_i) = a_i$, does no transformation:

$$M(a; f()) = \frac{1}{N} \sum f(a_i)$$

We'll also need a function to compute the standard deviation for a variable, a.

$$S(a) = \sqrt{\frac{\sum(a_i - \bar{a})^2}{N}}$$

This lets us define a number of related values as mean values after some transformation.

$$\bar{x} = M(x; f(a_i) = a_i)$$

$$\bar{y} = M(y; f(a_i) = a_i)$$

$$\overline{x^2} = M(x; f(a_i) = a_i^2)$$

$$\overline{y^2} = M(y; f(a_i) = a_i^2)$$

$$\overline{xy} = M(x, y; f(a_i, b_i) = a_i \times b_i)$$

From these individual values, we can compute the correlation coefficient, r_{xy}.

$$r_{xy} = \frac{\overline{xy} - \bar{x}\bar{y}}{\sqrt{(\overline{x^2} - \bar{x}^2)(\overline{y^2} - \bar{y}^2)}}$$

In addition to the above values, we need two more values for the standard deviations of the two variables.

$$s_x = S(x) s_y = S(y)$$

From the correlation coefficient, and the two standard deviations, we can compute the coefficient of the line, m, and the intercept value, b.

$$m = r_{xy} \frac{s_y}{s_x}$$

$$b = \bar{y} - m\bar{x}$$

This yields the coefficient, m, and intercept, b, for the equation $y = mx + b$, which minimizes the error between the given samples and the line. This is computed using one higher-order function, $M(a; f())$, and one ordinary function, $S(a)$. This doesn't seem to be a significant improvement over other methods. Because it's built using standard library functions and functional programming techniques, it can be applied to any Python data structure. This can save the step of transforming data into **numpy** array objects.

17

Next Steps

The journey from raw data to useful information has only begun. There are often many more steps to getting insights that can be used to support enterprise decision-making. From here, the reader needs to take the initiative to extend these projects, or consider other projects. Some readers will want to demonstrate their grasp of Python while others will go more deeply into the area of exploratory data analysis.

Python is used for so many different things that it seems difficult to even suggest a direction for deeper understanding of the language, the libraries, and the various ways Python is used.

In this chapter, we'll touch on a few more topics related to exploratory data analysis. The projects in this book are only a tiny fraction of the kinds of problems that need to be solved on a daily basis.

Every analyst needs to balance the time between understanding the enterprise data being processed, searching for better ways to model the data, and effective ways to present the results. Each of these areas is a large domain of knowledge and skills.

We'll start with a review of the architecture underlying the sequence of projects in this book.

Overall data wrangling

The applications and notebooks are designed around the following multi-stage architecture:

- Data acquisition

- Inspection of data

- Cleaning data; this includes validating, converting, standardizing, and saving intermediate results

- Summarizing, and the start of modeling data

- Creating deeper analysis and more sophisticated statistical models

The stages fit together as shown in *Figure 17.1.*

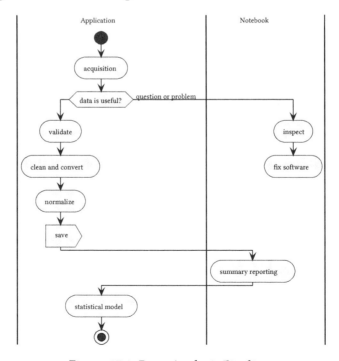

Figure 17.1: Data Analysis Pipeline

The last step in this pipeline isn't — of course — final. In many cases, the project evolves from exploration to monitoring and maintenance. There will be a long tail where the model continues to be confirmed. Some enterprise management oversight is an essential part of this ongoing confirmation.

In some cases, the long tail is interrupted by a change. This may be reflected by a model's inaccuracy. There may be a failure to pass basic statistical tests. Uncovering the change and the reasons for change is why enterprise management oversight is so essential to data analysis.

This long tail of analysis can last a long time. The responsibility may be passed from analyst to analyst. Stakeholders may come and go. An analyst often needs to spend precious time justifying an ongoing study that confirms the enterprise remains on course.

Other changes in enterprise processing or software will lead to outright failure in the analytical processing tools. The most notable changes are those to "upstream" applications. Sometimes these changes are new versions of software. Other times, the upstream changes are organizational in nature, and some of the foundational assumptions about the enterprise need to change. As the data sources change, the data acquisition part of this pipeline must also change. In some cases, the cleaning, validating, and standardizing must also change.

Because of the rapid pace of change in the supporting tools — Python, JupyterLab, Matplotlib, etc. — it becomes essential to rebuild and retest these analytic applications periodically. The version numbers in `requirements.txt` files must be checked against Anaconda distributions, conda-forge, and the PyPI index. The tempo and nature of the changes make this maintenance task an essential part of any well-engineered solution.

The idea of enterprise oversight and management involvement is sometimes dubbed "decision support." We'll look briefly at how data analysis and modeling is done as a service to decision-makers.

The concept of "decision support"

The core concept behind all data processing, including analytics and modeling, is to help some person make a decision. Ideally, a good decision will be based on sound data.

In many cases, decisions are made by software. Sometimes the decisions are simple rules that identify bad data, incomplete processes, or invalid actions. In other cases, the decisions are more nuanced, and we apply the term "artificial intelligence" to the software making the decision.

While many kinds of software applications make many automated decisions, a person is still — ultimately — responsible for those decisions being correct and consistent. This responsibility may be implemented as a person reviewing a periodic summary of decisions made.

This responsible stakeholder needs to understand the number and types of decisions being made by application software. They need to confirm the automated decisions reflect sound data as well as the stated policy, the governing principles of the enterprise, and any legal frameworks in which the enterprise operates.

This suggests a need for a meta-analysis and a higher level of decision support. The operational data is used to create a model that can make decisions. The results of the decisions become a dataset about the decision-making process; this is subject to analysis and modeling to confirm the proper behavior of the operational model.

In all cases, the ultimate consumer is the person who needs the data to decide if a process is working correctly or there are defects that need correction.

This idea of multiple levels of data processing leads to the idea of carefully tracking data sources to understand the meaning and any transformations applied to that data. We'll look at metadata topics, next.

Concept of metadata and provenance

The description of a dataset includes three important aspects:

- The syntax or physical format and logical layout of the data

- The semantics, or meaning, of the data

- The provenance, or the origin and transformations applied to the data

The physical format of a dataset is often summarized using the name of a well-known file format. For example, the data may be in CSV format. The order of columns in a CSV file may change, leading to a need to have headings or some metadata describing the logical layout of the columns within a CSV file.

Much of this information can be enumerated in JSON schema definitions.

In some cases, the metadata might be yet another CSV file that has column numbers, preferred data types, and column names. We might have a secondary CSV file that looks like the following example:

```
1,height,height in inches
2,weight,weight in pounds
3,price,price in dollars
```

This metadata information describes the contents of a separate CSV file with the relevant data in it. This can be transformed into a JSON schema to provide a uniform metadata notation.

The provenance metadata has a more complicated set of relationships. The PROV model (see `https://www.w3.org/TR/prov-overview/`) describes a model that includes **Entity**, **Agent**, and **Activity**, which create or influence the creation of data. Within the PROV model, there are a number of relationships, including **Generation** and **Derivation**, that have a direct impact on the data being analyzed.

There are several ways to serialize the information. The PROV-N standard provides a textual representation that's relatively easy to read. The PROV-O standard defines an OWL ontology that can be used to describe the provenance of data. Ontology tools can query the graph of relationships to help an analyst better understand the data being analyzed.

The reader is encouraged to look at `https://pypi.org/project/prov/` for a Python implementation of the PROV standard for describing data provenance.

In the next section, we'll look at additional data modeling and machine learning applications.

Next steps toward machine learning

We can draw a rough boundary between statistical modeling and machine learning. This is a hot topic of debate because — viewed from a suitable distance — all statistical modeling can be described as machine learning.

In this book, we've drawn a boundary to distinguish methods based on algorithms that are finite, definite, and effective. For example, the process of using the linear least squares technique to find a function that matches data is generally reproducible with an exact closed-form answer that doesn't require tuning hyperparameters.

Even within our narrow domain of "statistical modeling," we can encounter data sets for which linear least squares don't behave well. One notable assumption of the least squares estimates, for example, is that the independent variables are all known exactly. If the x values are subject to observational error, a more sophisticated approach is required.

 The boundary between "statistical modeling" and "machine learning" isn't a crisp, simple distinction.

We'll note one characteristic feature of machine learning: tuning hyperparameters. The exploration of hyperparameters can become a complex side topic for building a useful model. This feature is important because of the jump in the computing cost between a statistical model and a machine learning model that requires hyperparameter tuning.

Here are two points on a rough spectrum of computational costs:

- A statistical model may be created by a finite algorithm to reduce the data to a few parameters including the coefficients of a function that fits the data.

- A machine learning model may involve a search through alternative hyperparameter values to locate a combination that produces a model passes some statistical tests for utility.

The search through hyperparameter values often involves doing substantial computation to create each variation of a model. Then doing additional substantial computations to measure the accuracy and general utility of the model. These two steps are iterated for various hyperparameter values, looking for the best model. This iterative search can make some machine learning approaches computationally intensive.

This overhead and hyperparameter search is not a universal feature of machine learning. For the purposes of this book, it's where the author drew a line to limit the scope, complexity, and cost of the projects.

You are strongly encouraged to continue your projects by studying the various linear models available in scikit-learn. See `https://scikit-learn.org/stable/modules/linear_model.html`.

The sequence of projects in this book is the first step toward creating a useful understanding from raw data.

Subscribe to our online digital library for full access to over 7,000 books and videos, as well as industry leading tools to help you plan your personal development and advance your career. For more information, please visit our website.

Why subscribe?

- Spend less time learning and more time coding with practical eBooks and Videos from over 4,000 industry professionals
- Improve your learning with Skill Plans built especially for you
- Get a free eBook or video every month
- Fully searchable for easy access to vital information
- Copy and paste, print, and bookmark content

Did you know that Packt offers eBook versions of every book published, with PDF and ePub files available? You can upgrade to the eBook version at packt.com and as a print book customer, you are entitled to a discount on the eBook copy. Get in touch with us at customercare@packtpub.com for more details.

At www.packtpub.com, you can also read a collection of free technical articles, sign up for a range of free

Other Books You Might Enjoy

If you enjoyed this book, you may be interested in these other books by Packt:

<packt>

Causal Inference and Discovery in Python

Unlock the secrets of modern causal machine learning
with DoWhy, EconML, PyTorch and more

ALEKSANDER MOLAK

Causal Inference and Discovery in Python

Aleksander Molak

ISBN: 9781804612989

- Master the fundamental concepts of causal inference
- Decipher the mysteries of structural causal models
- Unleash the power of the 4-step causal inference process in Python
- Explore advanced uplift modeling techniques
- Unlock the secrets of modern causal discovery using Python
- Use causal inference for social impact and community benefit

Python for Geeks

Muhammad Asif

ISBN: 9781801070119

- Understand how to design and manage complex Python projects
- Strategize test-driven development (TDD) in Python
- Explore multithreading and multiprogramming in Python
- Use Python for data processing with Apache Spark and Google Cloud Platform (GCP)
- Deploy serverless programs on public clouds such as GCP
- Use Python to build web applications and application programming interfaces
- Apply Python for network automation and serverless functions
- Get to grips with Python for data analysis and machine learning

Python Data Analysis - Third Edition

Avinash Navlani, Armando Fandango, Ivan Idris

ISBN: 9781789955248

- Explore data science and its various process models
- Perform data manipulation using NumPy and pandas for aggregating, cleaning, and handling missing values
- Create interactive visualizations using Matplotlib, Seaborn, and Bokeh
- Retrieve, process, and store data in a wide range of formats
- Understand data preprocessing and feature engineering using pandas and scikit-learn
- Perform time series analysis and signal processing using sunspot cycle data
- Analyze textual data and image data to perform advanced analysis
- Get up to speed with parallel computing using Dask

Packt is searching for authors like you

If you're interested in becoming an author for Packt, please visit `authors.packtpub.com` and apply today. We have worked with thousands of developers and tech professionals, just like you, to help them share their insight with the global tech community. You can make a general application, apply for a specific hot topic that we are recruiting an author for, or submit your own idea.

Share your thoughts

Now you've finished *Python Real-World Projects*, we'd love to hear your thoughts! Scan the QR code below to go straight to the Amazon review page for this book and share your feedback or leave a review on the site that you purchased it from.

https://packt.link/r/1803246766

Your review is important to us and the tech community and will help us make sure we're delivering excellent quality content.

Download a free PDF copy of this book

Thanks for purchasing this book!

Do you like to read on the go but are unable to carry your print books everywhere? Is your eBook purchase not compatible with the device of your choice?

Don't worry, now with every Packt book, you get a DRM-free PDF version of that book at no cost.

Read anywhere, any place, on any device. Search, copy, and paste code from your favorite technical books directly into your application.

The perks don't stop there, you can get exclusive access to discounts, newsletters, and great free content in your inbox daily

Follow these simple steps to get the benefits:

1. Scan the QR code or visit the link below

https://packt.link/free-ebook/9781803246765

2. Submit your proof of purchase

3. That's it! We'll send your free PDF and other benefits to your email directly

Index

www.ingramcontent.com/pod-product-compliance
Lightning Source LLC
Chambersburg PA
CBHW060643060326
40690CB00020B/4501